RED BANDIT
The MiG-29 in Combat

Mike Guardia

Copyright 2025 © Mike Guardia

From *Debrief: A Complete History of US Aerial Engagements, 1981-present* by Craig Brown. Used by permission of Schiffer Publishing. Any third-party use of this material, outside of this publication, is prohibited. Interested parties must apply directly to Schiffer Publishing for permission.

From *Ethiopian-Eritrean Wars: Volume 2*, by Tom Cooper, Adrien Fontanellaz, Helios & Company, used in accordance with UK copyright laws regarding Fair Usage, quoting 800 words or less.

From *Mikoyan MiG-29 & MiG-35: Famous Russian Aircraft*, by Yefim Gordon and Dmitriy Komissarov, Crecy Publishing, used in accordance with UK copyright laws regarding Fair Usage, quoting 800 words or less.

From *Operation Allied Force - Volume 1: Air War Over Serbia, 1999*, by Bojan Dimitrijevic and Jovica Draganić, Helios & Company, used in accordance with UK copyright laws regarding Fair Usage, quoting 800 words or less.

From *War in Ukraine - Volume 6: The Air War February-March 2022*, by Tom Cooper, Adrien Fontanellaz, and Milos Sipos; Helios & Company, used in accordance with UK copyright laws regarding Fair Usage, quoting 800 words or less.

Published by Magnum Books
PO Box 1661
Maple Grove, MN 55311

www.mikeguardia.com

ISBN-13: 979-8-9917981-3-6

All rights reserved, including the right to reproduce this book or any part of this book in any form by any means including digitized forms that can be encoded, stored, and retrieved from any media including computer disks, CD-ROM, computer databases, and network servers without permission of the publisher except for quotes and small passages included in book reviews. Copies of this book may be purchased for educational, business or promotional use.

Also by Mike Guardia

The Combat Diaries
Hal Moore: A Soldier Once . . . and Always
Skybreak
Tomcat Fury
Danger Forward
Wings of Fire
Foxbat Tales
Fire in the Hole

Co-authored with LTG Harold G. Moore

Hal Moore on Leadership: Winning When Outgunned and Outmanned

A Soviet MiG-29 from the 31st Guards Fighter Regiment, forward-stationed in Falkenberg, East Germany. *Rob Schleiffert*

Another Soviet MiG-29 at the Farnborough Air Show in Hampshire, England, 1988. *Anthony Noble*

Table of Contents

Introduction   1

Chapter 1: Before the Dawn   5

Chapter 2: Breaking Ranks   27

Chapter 3: The Baghdad Express   35

Photo Section   57

Chapter 4: Balkanized   75

Chapter 5: Brushfire Wars & The Arab Spring   99

Chapter 6: The Road to Kyiv   131

Epilogue   161

Bibliography   165

Artist Richard J. Terry created this 1986 rendition of the MiG-29 escorting a Tu-22M strategic bomber. Commissioned by the US Department of Defense, this painting reflected the Soviet Union's drive to modernize its air forces in the mid-1980s. *Defense Intelligence Agency*

Introduction

Few fighter jets have experienced the same range of combat scenarios, technological evolutions, and geopolitical turbulence as the MiG-29. Designed during the height of the Cold War, the MiG-29 (NATO reporting name: "Fulcrum") was built for speed, agility, and air superiority. However, as history unfolded, the aircraft found itself in battles far removed from the world that its Soviet designers had envisioned.

From its first engagements in the 1990s to its latest deployments in the 21st Century, the MiG-29 has played a critical role in numerous conflicts from Europe to the Middle East. Like many of its Mikoyan predecessors, the MiG-29 was exported to the far reaches of the Arab World and the Communist Bloc, with each country adapting it to their own needs and combat doctrines. Some air forces have pushed the aircraft to its limits, modernizing it with newer avionics and upgraded weapons, while others have struggled to maintain their fleets due to economic constraints and geopolitical upheavals. Despite these challenges, the MiG-29 has remained relevant, proving itself as a capable fighter in both air-to-air and ground-attack roles.

Soviet doctrine envisioned the MiG-29 as a frontline interceptor, operating alongside the heavier Sukhoi Su-27 to create a layered air defense network. Following its maiden flight in 1977, the MiG-29 entered service in the mid-1980s – sleek, fast, and branded as the new apex of modern airpower.

Serving among the various Eastern and Eurasian air forces, the MiG-29 made its combat debut in the skies over Iraq during Operation Desert Storm. It saw significant action in later conflicts such as the Kosovo War, the Kargil War, and the Eritrean-Ethiopian Border War. Some of the MiG-29's earliest engagements yielded mixed results – highlighting both its impressive dogfighting capabilities and the vulnerabilities of outdated avionics, training disparities, and lack of maintenance assets.

However, the most defining moment in the MiG-29's history has been its role in the ongoing Russia-Ukraine War. Both the Russian and Ukrainian Air Forces have deployed the MiG-29, with Ukrainian pilots using the jet to remarkable effect against a numerically and technologically-superior opponent. Now modified to carry Western-built missiles, including the AGM-88, the aircraft has been repurposed in ways its original designers could never have imagined. It has intercepted cruise missiles and drones, flown daring ground-attack missions, and played a key role in Ukraine's efforts to reclaim its airspace from Russian aggression.

Red Bandit explores the combat history of the MiG-29, tracing its journey from a Soviet-era interceptor and defensive fighter to the battle-hardened, multi-role aircraft still flying in the world's most volatile regions. Through firsthand accounts, detailed battle analyses, and technical evaluations, this book examines how the Fulcrum has shaped aerial warfare and how it continues to influence modern air combat.

Whether guarding the Soviet frontier, patrolling the former Yugoslavia, or flying intercepts over Kyiv, the MiG-29's story is one of resilience, adaptability, and an enduring legacy in the realm of combat aviation.

A Soviet MiG-29 over Alaska en route to the Abbotsford Air Show in British Columbia, Canada, August 1989. *US Air Force*

The same MiG-29 sits parked on the ramp following its demonstration flight at the Abbotsford Air Show. *US Air Force*

Chapter 1:
Before the Dawn

He who sees farther and shoots farther enjoys an undisputed advantage over the adversary; that's the law [of aerial combat]. It takes an Anti-F-15 to oppose the F-15; and that's the aircraft we need most. – Air Marshal A.N. Yefimov

Air Marshal Aleksandr N. Yefimov described it best when he articulated the need for an "anti F-15." Yefimov knew, as did many within the Soviet Air Forces, that a true counterbalance to the F-15 would secure the Soviets' parity with the oncoming wave of fourth-gen fighters.

By the end of the 1960s, the Soviet Defense Ministry had taken note of the events in Vietnam. The MiG-17 and MiG-21 had performed well during the early years of the conflict, but both fighters were now falling prey to the F-4 Phantom II. The newer MiG-23 and MiG-25 were on the horizon, with the former featuring a variable-geometry "swing-wing" design, but both aircraft were still third-generation stock. Moreover, neither of these newer MiGs had been optimized for air-to-air combat.

Thus, in September 1969, the Soviet Air Forces and the Ministry of Aircraft Industry met to discuss the requirements for a next-generation fighter. Officially dubbed the "Future Tactical Fighter"

(*Perspektivnyy Frontovy Istrebitel* – or "PFI") program, this forthcoming fighter had to achieve parity with the best of NATO's frontline air squadrons.

To counter this long list of enemy aircraft, the PFI-specification fighter would have to accomplish the following:

- Engage enemy fighters from beyond visual range and in close-quarters, air-to-air combat;
- Escort friendly bombers, transports, and reconnaissance aircraft;
- Destroy all classes of enemy aircraft;
- Intercept aerial targets at long ranges;
- Perform aerial reconnaissance;
- Provide tactical air cover, and;
- Perform select ground attack missions;

To fulfill these mission parameters, the PFI would need to have a long endurance; a short takeoff capability; pulse Doppler radar; speeds in excess of Mach 2; a complement of medium- and short-range missiles; and a 30mm autocannon.

By 1972, however, Soviet military planners decided to split the PFI into *two* programs – one for a "lightweight" variant; the other for a "heavyweight" variant. Analyses of Western hardware and tactics revealed that a single fighter could not handle the broad range of missions that had emerged on the modern battlefield. For example, an interceptor needed to be fast and heavily-armed; but heavy arms and fuel requirements would sacrifice its desired agility. Reconciling these requirements, Soviet defense ministers assigned the "heavy" program to Sukhoi, while the "light" variant fell to the Mikoyan Design Bureau. To differentiate the now-parallel programs, Sukhoi's project retained the legacy "PFI" designation, while Mikoyan's became the "LPFI" (*Perspektivnyy Lyogkiy Frontovoy Istrebitel*), or "Advanced Lightweight Tactical Fighter." Within these design parameters, the Soviet fighter fleet would have approximately 33% heavy aircraft (PFI) and 67% lightweight aircraft (LPFI). Respectively, the PFI and LPFI designs would evolve into the Su-27 and MiG-29 – both of which would define the Soviet fighter force of the 1980s.

By the time Mikoyan had taken the reins of the LPFI program, the design bureau had been a longstanding powerhouse in the Soviet aerospace industry. Each of their aircraft carried the prefix "MiG" (referencing the original *Mikoyan and Gurevich* design team) and the term itself had become synonymous with Soviet jet fighters. In fact, the MiG-15 had given rise to the popular term "MiG Alley" during the Korean War.

In its definitive form, the MiG-29 featured a blended wing body (BWB) design, punctuated by twin tails and differentially-movable stabilizers. For its powerplant, Mikoyan settled on the twin-mounted Klimov RD-33 engine suite. Rated at more than 80 kilonewtons in afterburner, the RD-33 could easily send the MiG to speeds in excess of Mach 2. Wide spacing between the two engines facilitated greater lift and reduced wing loading, subsequently improving the plane's maneuverability. However, due to its relatively short combustor, the RD-33 produced a noticeably heavier smoke trail than many of its contemporaries.

The BWB design, however, gave the airframe a significant amount of structural resiliency, enabling the MiG to withstand high-G loads and maneuver at high angles of attack. Mikoyan had also designed the plane with a high concentration of composite and aluminum-lithium alloys to reduce the airframe's weight.

With its aerodynamically-refined body and capable powerplant, the MiG-29 had a baseline range of 930 miles (1,500 kilometers). With external fuel tanks, however, that range increased to 1,300 miles (2,100 kilometers).

Internal fuel capacity aboard the first production-model MiG-29s (designated "9.12;" NATO Reporting Name: "Fulcrum-A") was more than 4,000 liters, distributed across six internal fuel tanks: four in the fuselage; one in each wing. Successive variants of the MiG-29, however, were built with an increased fuel capacity. The Fulcrum-C, for instance, could carry some 4,540 liters, thanks to a larger primary fuselage tank. For longer flights, the Fulcrum-A could accommodate a 1,500-liter centerline drop tank, while later production variants carried two 1,150-liter underwing drop tanks.

But even with its rated fuel capacity and mission endurance, the MiG-29 could never be a true "expeditionary" fighter like its Su-27 stablemate. In fact, with a full mission payload, the earliest Fulcrum-A models could barely reach a 60-mile (100-kilometer) radius from any given airfield. Thus, even during the prototype and development phase, Soviet planners knew that the MiG-29 would be tethered to a Defensive Counter Air (DCA) strategy – operating mostly over friendly territory, interdicting aerial attacks against friendly targets.

Still, the MiG-29 was better-armed, better-equipped, and more aerodynamic than the MiG-23 Flogger – which, at the time, was the backbone of the Soviet Union's fighter fleet. Early production variants of the MiG-29 featured the GSh-30-1 30mm autocannon with a 150-round magazine. Later variants, however, reduced the gun to a 100-round capacity, limiting the pilot's trigger time to a few multi-round bursts.

The missile suite, however, was more promising. Indeed, its pylons could accommodate the R-27, R-60, and R-73 air-to-air missiles.

Inside the cockpit, the pilot sat in an elevated position, encased in a bubble canopy offering unrestricted all-round visibility, while sitting atop a K-36DM ejection seat.

During its initial production run, the MiG-29 featured some of the most advanced avionics in the Soviet Union. For example, the MiG-29 "9.12" base model carried the Phazotron RLPK-29 radar fire control system. This fire control suite included the Sapfir-29 pulse-Doppler radar (with a true "look-down/shoot-down" capability), and the Ts100 digital computer.

The Sapfir-29 was an updated version of the Sapfir-23ML used aboard latter-day variants of the MiG-23. The Soviet Defense Ministry had instructed Phazotron to produce a modern radar for the MiG-29. According to specifications, the new radar was supposed to have a flat planar array antenna and full digital signal processing, for tracking aerial targets at distances up to 60 miles (100 kilometers).

However, during prototype testing, Phazotron's engineering team realized they couldn't build such a complex radar within the requisite timeframe and still make it small enough to fit within the MiG-29's nose. Thus, rather than design a new radar, Phazotron simply retooled the Sapfir-23ML's mechanisms, paired it to a new Ts100 digital computer, and called it "Sapfir-29." To be fair, the slapdash Sapfir-29 was a fully-functional radar, but it inherited all the weaknesses of its predecessor. Consequently, this radar hindered the MiG-29's ability to detect and track targets at ranges compatible with the newer R-27 and R-77 missiles.

By the late 1980s, however, the Soviets had developed the new N019M Topaz radar for the upgraded MiG-29S variants. Although the Topaz was a marked improvement over the Sapfir system, it still fell short of expectations for maintaining airborne situational awareness.

Finally, with the onset of the MiG-29M variant in the 1990s-2000s, the modern-day Russian Air Force received an onboard radar that met the original specifications of their Soviet forbearers. The N010 Zhuk-M radar boasted a planar array antenna and superior processing, along with multi-target engagement capability and full compatibility with the newer R-77 missile.

After completing its maiden flight in October 1977, the MiG-29 formally entered Soviet service in 1983. By the time of its unveiling, however, the West had known of the MiG-29's existence since early 1979, when US satellite imagery revealed one of the Mikoyan prototypes at an airfield in Zhukovskiy. Of course, Western analysts couldn't determine the aircraft's manufacturer from the grainy satellite photos, so they gave the mystery fighter a provisional reporting name: "Ram-L," which by itself was erroneous because the US had mistakenly identified Zhukovskiy airfield as "Ramenskoye" airfield (referring to the name of a nearby town).

By 1982, when NATO learned that the Ram-L was a Mikoyan fighter, they christened it with the reporting name "Fulcrum." Although the Soviets had no internal naming conventions for their own planes, Soviet pilots reportedly liked the nickname "Fulcrum" so much that they began using it within their own doctrinal literature.

Still, it was a long time before NATO fully knew what the MiG-29 looked like. Artistic renditions published in 1982 were wildly inaccurate, showing a distorted MiG-29 that bore a stronger resemblance to the American F/A-18. In fact, Western analysts didn't get their first sustained look at the elusive Fulcrum until *July 1986*, when the Soviets put a MiG-29 on public display at a Finnish air show.

Two years later, the Soviet Air Forces sent a pair of MiG-29s for display at the Farnborough Airshow in the United Kingdom. The following year, a MiG-29 participated in aerial maneuvers at the 1989 Paris Air Show. Although the plane suffered a non-fatal crash during the first week of that airshow, Western observers were still impressed by the Fulcrum's speed and agility.

The Soviet Union, naturally, was the first and largest operator of the MiG-29. The first production-model variant (the aforesaid "9.12," or "Fulcrum-A") gradually filtered into frontline service with the Soviet Air Forces from 1983-86. The forward-stationed units in Eastern Europe received more than 350 MiG-29s by mid-decade. For example, the Western Group of Forces (stationed in East Germany) received 249 Fulcrums, while the Southern Group of Forces (Hungary), operated 101 MiG-29s. Soviet air units in the Central Group of Forces (Czechoslovakia), however, received only ten. Within the borders of the

Soviet homeland, 348 MiG-29s were stationed in the western republics, with an additional 117 Fulcrums based in Central Asia and the Far Eastern regions of the USSR.

By 1986, an updated version of the base model MiG-29 (designated "9.13," or "Fulcrum-C") entered service. The upgraded MiG had a better fuel capacity and came equipped with the new L-203BE Gardenyia-1 jammer. The MiG-29S variant soon followed, retaining the "Fulcrum-C" NATO designation, but featuring the aforesaid Topaz radar. By 1991, the latter-day Soviet Air Forces had more than 880 MiG-29s in service.

After the collapse of the USSR, the former Soviet MiGs were passed on to the various successor states. The Russian Air Force, for example, retained 580 MiG-29s, including the R&D aircraft owned by Mikoyan. By 2002, however, this number had fallen to nearly 490.

But while the legacy era MiG-29s dwindled in number, Mikoyan and the Russian Air Force accelerated development of the MiG-29M and MiG-29K variants. The MiG-29M (Fulcrum-E) is a specially-modified multirole variant of the legacy plane, with accommodations for air-to-air combat *and* ground attack missions. The Fulcrum-E features a redesigned airframe, fly-by-wire controls, and an upgraded powerplant. The MiG-29K (Fulcrum-D) is the carrier-based variant. To accommodate its maritime mission, the MiG-29K features folding wings and the normal variety of arrestor gear. Additional upgrades included a radar-absorbent exterior coating and improved cockpit displays paired to a *Topsight E* helmet-mounted targeting system.

At this writing, the MiG-29 remains in service with the Russian Air Force (~250 planes according to most sources). However, Russia's MiG-29 fleet is steadily being replaced by the growing number of Sukhoi "Flanker" jets and the newer MiG-35 – which, itself, is an upgraded version of the MiG-29K.

Among the former Soviet republics, Belarus and Ukraine became the largest operators of MiG-29s outside the Russian Federation. In 1992, Belarus acquired more than 84 MiG-29s (including Fulcrum-A/Cs and two-seat trainer variants) from the former Soviet 927th Fighter Aviation Regiment at Beryozy Air Base.

Today, however, only 33 MiG-29s remain in service with the Belarusian Air Force, although more than a dozen have been upgraded to the Belarusian-exclusive "MiG-29BM" strike fighter variant. This domestically-produced "BM" package includes an upgraded weapons suite, modern radar, and air-to-air refueling capabilities.

Ukraine, meanwhile, received the lion's share of post-Soviet Fulcrums. By late 1992, the newly-formed Ukrainian Air Force had taken possession of all ex-Soviet MiGs within their borders, as well as some of the Fulcrums being withdrawn from the now-reunified Germany. In total, Ukraine amassed more than 200 MiG-29s, including 155 Fulcrum-C variants.[1]

Kazakhstan, meanwhile, took over the 715th Fighter Aviation Regiment at Loogovaya Air Base. This unit included twelve standard Fulcrums and two trainer variant MiG-29UBs. In 1995, the Kazakh Air Defense Forces took delivery of twenty-two additional MiG-29s from Russia (eighteen Fulcrum-A variants and four trainers).

At this writing, however, the current status of Kazakhstan's MiG-29 fleet remains unclear. As of 2023, the Kazakh Air Defense Forces had nearly two dozen MiG-29s in service. However, in October of that year, reports surfaced that Kazakhstan had retired its fleet of older MiGs and had put them up for auction. It was later reported that the US purchased these planes in April 2024, with the intent of transferring them to Ukraine for spare parts usage in the ongoing Russia-Ukraine War. Kazakhstan has denied these claims; and the Pentagon has yet to provide comment. Meanwhile, according to data from the International Institute for Strategic Studies, Kazakhstan's inventory of MiG-29s in 2024 stood at twelve planes – indicating a downsized fleet, but not yet retired.

Although Kazakhstan's fleet of MiG-29s may or may not be extinct, there is no question that the

---

[1] See Chapter 6, "The Road to Kyiv," for a more detailed discussion on the history of Ukrainian MiG-29s.

Moldovan Air Force has long since retired the last of their Fulcrums. After the collapse of the Soviet Union, Moldova inherited 33 MiG-29s based at Markuleshty – home to the 86th Guards Maritime Fighter Aviation Regiment. However, the newly-formed Moldovan government soon realized that the MiG-29 was too costly for its annual defense budget. Eventually, Yemen and Romania purchased a handful of these Moldovan MiGs. Their biggest sale, however, came from an unlikely customer – the United States. Towards the end of the 1990s, Moldova offered to sell their remaining Fulcrums to the Iranian Air Force, which had been operating the fighter since 1989. At some point, the US was alerted to the pending Iranian sale, and stepped in with an offer of $80 million (USD) to purchase 21 Fulcrum-Cs and 500 air-to-air missiles from Moldova. Because the Fulcrum-C had tactical nuclear strike capabilities, the American government hastily purchased the fighters so they wouldn't end up in the hands of "rogue states" like Iran. As of 2024, Moldova has six MiG-29s in storage, none of which are operational.

Turkmenistan and Uzbekistan inherited sizeable quantities of the MiG-29 after the fall of the Soviet Union, but their air forces have never provided many details regarding their daily operations. At this writing, it's known that the Turkmen Air Force operates nearly two dozen MiG-29s, most of which (if not all) are the very same Soviet-era crates they inherited in 1992. Uzbekistan, meanwhile, drew many of its present-day MiG-29s from the former 115th Fighter Aviation Regiment at Kokaidy Air Base, close to the Afghan border. Reports vary as to the number of Uzbek MiG-29s in service, but most sources place the number between 30-38.

Azerbaijan, on the other hand, had no MiG-29 units within its borders when the Soviet Union collapsed. In fact, it would be more than a decade until this former Soviet republic would see its first Fulcrums. Azerbaijan began soliciting its regional neighbors for MiG-29 sales during the early 2000s. They intended to phase out their older MiG-25s as part of their modernization efforts for the Azerbaijani Air Force. Ironically, Ukraine determined that its own air force had a surplus of MiG-29s inherited from the Soviet Union. Thus, Ukraine transferred nearly two dozen of its "excess" MiGs to Azerbaijan between 2006-2008.

These MiG-29s were purportedly upgraded before delivery – retooled with advanced avionics and the latest weaponry. Today, Azerbaijan's MiG-29 fleet remains an integral part of its air force, with fourteen active airframes in service. Unlike many former Soviet republics, however, Azerbaijan intends to keep the MiG-29 as its primary fighter for as long as possible.

Beyond the Soviet republics, the primary Eastern Bloc recipients of the MiG-29 included Yugoslavia, Czechoslovakia, East Germany, Hungary, Poland,

Romania, and Bulgaria.[2] The latter-day country of Czechoslovakia took delivery of its first Fulcrums in the spring of 1989. The previous year, a cadre of fifteen carefully-vetted Czechoslovakian pilots had been selected for Fulcrum training at Loogovaya Air Base in the Soviet Union. The Czechoslovak 11th Fighter Regiment, based at Zatec, was the first to receive these new MiG-29s. Ultimately, the Czechoslovak Air Force received nearly two dozen Fulcrums before the country dissolved.

Indeed, by 1992, the so-called "Velvet Revolution" had reached critical mass, calling for a peaceful partition between the Czechs and Slovaks. The resulting "Velvet Divorce," as it came to be known, quietly partitioned the Federal Republic of Czechoslovakia into the modern-day Czech Republic and Slovakia.

Naturally, this partition also divided the Czechoslovak Air Force's assets between the two successor states. For the MiG-29s, the odd-numbered aircraft went to Slovakia, while the even-numbered Fulcrums remained in the Czech Republic. By 1994, however, the Czechs were eager to join NATO. And they were just as eager to rid themselves of their Soviet-era hardware. Thus, in December 1995, the Czech government negotiated a deal to transfer all ten of their MiG-29s to the Polish Air Force in

---

[2] See Chapter 4, "Balkanized" for a more detailed discussion on the history of Yugoslav MiG-29s.

exchange for the new PZL W-3 Sokol helicopters.

Slovakia, meanwhile, held on to its MiG-29s well into the 21st Century. Aside from the handful of Fulcrums inherited from the Czechoslovak Air Force, Slovakia accepted twelve additional MiG-29s from Russia as compensation for prior debts. By 2009, most of these Slovakian MiGs had been upgraded to align with NATO standards.

However, on August 31, 2022, Slovakia formally withdrew the MiG-29 from service. Soon thereafter, Slovakian Foreign Minister Rastislav Káčer confirmed that his government would be transferring their entire fleet of MiG-29s to Ukraine. In April 2023, Slovakia delivered the last of its thirteen MiG-29s to the Ukrainian Air Force.

The German Democratic Republic (East Germany) was the first of the former Warsaw Pact countries to receive the MiG-29. Between 1988-89, the East German *Luftstreitkräfte* received twenty-four MiG-29s. After German reunification in 1990, these MiG-29s were absorbed into the West German *Luftwaffe*.

Under the reunified German banner, these post-Cold War MiGs had a unique opportunity to square off against Western fighters in mock dogfights during NATO's annual air exercises. The outcome of these aerial maneuvers yielded what many analysts have considered to be the fairest assessment of the Fulcrum's strengths and weaknesses. The German pilots noted that the MiG-29 was more maneuverable at slower speeds than the F-14 Tomcat, F-15 Eagle, and F-16 Fighting Falcon. Some pilots argued that the R-73 missile was better than the AIM-9 Sidewinder; but the German pilots were unanimous in their disdain for the R-27's user interface. Specifically, it hindered the pilot's ability to lock on to a target. German pilots also conceded that American fighters were superior during nighttime operations and adverse weather conditions. In the *Luftwaffe*'s final assessment, the MiG-29 was best utilized as a "point defense interceptor" – flying over friendly skies, protecting cities and other military installations. Offensive Counter Air (OCA) missions, however, including fighter sweeps over enemy airspace, were better left to other aircraft.

From 1993 onward, these German MiG-29s were stationed at Rostock–Laage Airport near the northern coast of the former East Germany. By 2003, however, the *Luftwaffe*, like many of their allies within the former Eastern Bloc, had grown tired of their Soviet-era equipment. Germany sold its remaining Fulcrums to the Polish Air Force in September 2003, retaining only one MiG-29 to put on permanent display at Laage.

Comparatively speaking, Romania and Bulgaria were late to receive their Cold War deliveries of the MiG-29 – arriving in 1989 and 1990, respectively. In Bulgaria's case, their year-end stocks for 1990 stood at eighteen MiG-29s. Years later, Bulgaria entered negotiations with Mikoyan and the Russian Defense Ministry to purchase upgraded variants of the

Fulcrum. Ultimately, Bulgarian President Petr Stoyanov refused the deal for economic reasons; opting instead to modernize and upgrade their existing fleet. As of 2022, Bulgaria has fourteen MiG-29s in service.

In the fall of 1989, the Romanian Air Force took delivery of thirty MiG-29s. Ironically, the final aircraft arrived in Romania just days before the revolution that ended the reign of Nicolae Ceausescu. After the fall of Communism, the Romanian Air Force maintained their fleet of Soviet-era MiG-29s before retiring them in 2003.

Ironically, the Hungarian Air Force didn't receive their MiG-29s until *after* the Cold War ended. For an order placed in 1988, their Fulcrums didn't arrive until 1993. Hungary maintained an active fleet of more than two dozen MiG-29s well into the early 2000s. By the mid-2010s, however, less than a handful of these MiGs remained airworthy. As such, the Hungarian Air Force began replacing their MiG-29s with Saab JAS 39 fighters, but intended to keep the Fulcrums in a reserve status. By 2023, however, Hungary had finally retired the last of its MiG-29s.

In the wake of Solidarity, and in the dying days of the Cold War, the Polish Air Force ordered twelve MiG-29s, delivered between 1989-90. After the Warsaw Pact formally dissolved in 1991, Poland began to make its bid for NATO membership. However, this meant that the Polish armed forces would have to comply with NATO standards.

At first, Poland intended to phase out its Soviet-era equipment in favor of Western hardware. At first glance, this was the easiest way to ensure interoperability with their new NATO allies. However, the Polish government soon realized that it could not yet afford to scrap the MiG-29 and re-arm their squadrons with expensive Western fighters.

Thus, the Polish Air Force had little choice but to maintain (and upgrade) their current fleet of MiG-29s. The Czechs' aforementioned trade offer for the W-3 helicopters was perfectly-timed, bringing the total number of MiG-29s in Polish service to twenty-two, and during a time when Poland needed to bolster the morale of its post-Cold War citizens.

After purchasing twenty-two additional Fulcrums from the *Luftwaffe* in 2003, more than a dozen were overhauled and taken into service. The rest were retained for spare parts, intended for cannibalization to keep the other Polish MiGs flying. By the end of the decade, Poland was the largest NATO operator of MiG-29s. Those numbers decreased, however, with the onset of the War in Ukraine. In March 2023, Polish President Andrzej Duda announced the transfer of four operational MiG-29s to Ukraine, with additional aircraft to be delivered after routine servicing. This made Poland the first NATO member to provide Ukrainian forces with fighter aircraft.

Beyond the Iron Curtain, and at the farthest edge of the Communist world, Cuba and North Korea also took delivery of the MiG-29. Both countries had been longstanding clients of the Soviet Union, and both had been regular recipients of the latest MiG-series fighters.

In Castro's Cuba, the first MiG-29s arrived at San Antonio de los Baños Air Base in October 1989. As of 2010, the Cuban Air Force has only one squadron of MiG-29s. At this writing, however, only four of these aircraft are known to be flyable.

North Korea purchased the MiG-29 in response to the ROK's acquisition of the F-16 Fighting Falcon. North Korea was unique among the MiG-29's foreign operators because it was the only country to obtain manufacturing rights to build their own domestic copies of the Fulcrum. North Korea was also the first buyer outside the former Soviet republics to obtain the new Fulcrum-C.

At its peak, the North Korean Air Force had more than twenty-six operational MiG-29s. Today, that number stands at around eighteen, but the Kim Dynasty Fulcrums nevertheless remain in service with the 55th Air Regiment of the 1st Air Combat Division at Sunchon Air Base (although some sources have placed these MiGs with the 57th Fighter Regiment at Onchon).

As with many of Mikoyan's aircraft, the Fulcrum found a reliable export market in North Africa, the Middle East, and the Subcontinent. Algeria, Egypt, Ethiopia, and Eritrea currently maintain fleets of MiG-

29s, albeit in various states of operability. The Chadian Air Force flew a handful of MiG-29s imported from Ukraine in 2014, but none are known to be flyable as of 2024.

During the late 1980s, Iran and Iraq made near-simultaneous purchases of the MiG-29 towards the end of the Iran-Iraq War. Although Iraq has long since retired its fleet of MiG-29s, the Iranian Air Force still maintains a modest fleet of nineteen Fulcrums. Syria received its first contingent of MiG-29s in 1988 and kept the plane in operation until December 2024, when the Syrian Air Force was destroyed following the final collapse of the Al-Assad regime.

The former People's Democratic Republic of Yemen (South Yemen) had been a long-standing recipient of Soviet military hardware. In 1990, the Yemen Arab Republic (North Yemen) merged with South Yemen to create the "Republic of Yemen." The new Yemeni Air Force absorbed the latter-day South Yemeni Fulcrums, and imported additional MiG-29s from Russia and Moldova throughout the 1990s.

However, with the onset of the Yemeni Civil War in 2014, and the ensuing Saudi-led intervention, most of Yemen's air power was destroyed on the ground. On paper, Yemen has nearly two dozen active MiG-29s, but the true status of their operability – or even their existence – remains unknown.

The Indian Air Force took delivery of its first MiG-29s in 1987, with upgrades and additional purchases made throughout the 1990s and 2000s. In February

2010, the Indian Naval Air Arm received its first contingent of MiG-29K carrier-based fighters. Today, India is the only country outside of Russia to operate the naval variant of the MiG-29. By 2021, the Indian Air Force had more than 60 MiG-29s in service, while the Indian Navy reported nearly 45 MiG-29K fighters in the fleet.[3]

Bangladesh, Mongolia, and Myanmar each currently maintain a modest fleet of MiG-29s. In June 1999, the Bangladesh Air Force ordered eight MiG-29s at a collective price of $115 million (USD) signed by the Chief of the Bangladesh Air Staff with an additional $9 million for training and logistical services. Ten Bangladeshi pilots and seventy technicians were selected for three months of training in Russia. The first four Bangladeshi MiGs were delivered in December 1999, with the other four arriving in February 2000. Each of the Bangladeshi Fulcrums entered service with the No.5 Squadron at Bashar Air Base near Dhaka.

The Myanmar Air Force expressed interest in acquiring the MiG-29 as early as 1996. They eventually placed an order in 2001 for nearly a dozen aircraft, including the Fulcrum-A and the two-seat trainer. The Burmese MiGs entered service in 2003, followed by an additional order of twenty MiGs in 2009, part of a $570 million defense package.

---

[3] See Chapter 5, "Brushfire Wars & The Arab Spring" for a more detailed discussion on the history of Indian MiGs.

Mongolia, eager to rebuild the fixed-wing capabilities of its air force, took delivery of six MiG-29UB two-seat variants. Throughout the Cold War, the Mongolian Air Force had made extensive use of the MiG-17 and MiG-21. After the fall of the Soviet Union, however, the number of serviceable MiG-21s gradually fell until 2010, after which the Mongolian Air Force operated only helicopters. But in November 2019, Russia surprisingly donated a pair of MiG-29UBs, with an additional four delivered sometime prior to 2021. Of the six MiG-29s in Mongolian service, however, only two are known to be flyable.

Russian MiG-29K, the naval variant of the Fulcrum, at the International Aviation and Space Show (MAKS) in Zhukovsky, Russia, 2007. *Dmitriy Pichugin*

Chapter 2: Breaking Ranks

By all outward appearances, Captain Alexander Zuyev was a model Soviet citizen. Born on July 17, 1961, he was a devout Communist and a veritable "Top Gun" within the ranks of the Soviet Air Forces. To the casual observer, no one would have suspected that he would become the last major defector of the Cold War. For on May 20, 1989, Zuyev flew his MiG-29 from the Mikha Tskhakaya Airbase in Soviet Georgia to Turkey, and defected to the United States.

Zuyev wasn't the first Soviet airman to escape to the West, but he was certainly among the most high-profile cases. More than a decade earlier, in 1976, Lieutenant Viktor Belenko defected from the Soviet Air Defense Forces, flying his MiG-25 to Hokkaido, Japan with the intent of seeking asylum in the US. The impounded MiG-25 became an intelligence bonanza for Western analysts as they deconstructed the plane to study its strengths and weaknesses. Alexander Zuyev tacitly hoped to repeat that exposé with the MiG-29.

As it turned out, Zuyev's disillusionment had followed a similar path to Belenko's. Throughout the 1980s, Zuyev had slowly been exposed to the failings of Soviet Communism. He discovered that the senior military leaders and party bosses were thugs, criminals, and petty tyrants whose actions belied the lofty ideals of Leninism. These same leaders, he said, played an ongoing game of deception and misinformation – both amongst themselves and with the Soviet people. "I was living in the country that lied to me," he later said, "about justice [and] about all the history of the Soviet Union."

According to Zuyev, one of his galvanizing events was the coverup following the shootdown of Korean Air Lines (KAL) Flight 007 in 1983. That event, which *Newsweek* had labeled "Murder in the Air," occurred when a KAL Boeing 747 flying from New York to Seoul accidentally strayed from its flight plan, and vectored into Soviet airspace over the Kamchatka Peninsula. In turn, the Soviet Air Defense Command scrambled a flight of Su-15s to intercept, one of which shot down the 747 as it vectored towards the Sea of Japan.

"Ten days before that flight," said Zuyev, "an arctic gale knocked out the radars on the Kamchatka Peninsula." These were the early warning radars to detect any intrusions into Soviet airspace. "They were not operational; and Moscow knew about this fact," he continued. Thus, it wasn't long before the defense commissars began to pressure the local station commanders to fix the radars. "They fixed some of those radars," said Zuyev, "but not all of them…so they didn't have full coverage of that airspace." This led to critical gaps in their visibility over Kamchatka's airspace. Nevertheless, the Soviet

Far East Command falsely reported to Moscow that their radars had been fixed. But if the radars had been working as they should, Soviet Air Defense Forces would have identified KAL 007 with enough lead time to prevent the tragic shootdown. The KAL incident, along with the Chernobyl disaster of 1986 and the Tbilisi massacre in 1989, finally convinced Zuyev that he had to escape.

Although the Western media has often portrayed Zuyev as an altruistic hero who freed himself from the bonds of a Communist regime, other sources (primarily Russian) have not portrayed him in a flattering light. No one disputes that he was an excellent pilot. But some sources have characterized him as arrogant, selfish, and undisciplined. According to Russian aerospace historian Yefim Gordon: "Zuyev had been suspended from active [flight] duty some time before the incident; his track record was so poor that his superiors were considering a dishonorable discharge." This narrative of Zuyev suggests that his arrogance and petty resentment were the true catalysts for his defection.

But whether Zuyev was acting out of moral conviction or spiteful treachery, he nonetheless arrived at the Mikha Tskhakaya Airbase on May 20, 1989 with a carefully-conceived plan. The previous day, he had baked a cake and laced it with sleeping pills. Upon arriving at his unit, the 176th Fighter Aviation Regiment, Zuyev announced that his wife was pregnant and that he had baked the cake for an impromptu celebration with his comrades. In reality, however, Zuyev was affecting his own escape by *drugging* his fellow pilots.

After sedating the pilots and ground crewmen of the Quick Reaction Alert (QRA) flight, Zuyev cut the phone lines to prevent anyone from alerting base security. He then walked to the flight line and approached the QRA MiG-29 designated "44 Blue." But in order to take flight, he would have to get beyond the two mechanics standing guard by the MiG. As in most air forces, QRA planes were guarded by ground crewmen who doubled as armed sentries to protect the planes from saboteurs.

Zuyev had already sedated the second-shift guardsmen back at the QRA building. Thus, he waited until the two mechanics on guard duty noticed that their reliefs were late for the shift change. When one of the mechanics went back to the squadron to see what was causing the delay, Zuyev made his move: He approached the remaining mechanic; told him that his replacement was running late; and Zuyev offered to guard the plane himself. The mechanic, already vexed by the time delay, happily handed Zuyev his rifle and walked away.

Meanwhile, the other mechanic had found everyone asleep at the QRA building, and immediately grew suspicious. Running back to the MiG, he confronted Zuyev.

But Zuyev was certain that he could overpower the

guard.

"In school, I was a wrestler," he said – confident that his Greco-Roman grappling skills could subdue the mechanic.

But little did Zuyev know that this unseeming mechanic had been a wrestler, too.

Thus, when he tried to disarm the mechanic, a struggle ensued.

During their brawl, the mechanic broke Zuyev's right thumb and smashed his head on the concrete. At one point during the altercation, both men reached for their guns. Most accounts agree that both men fired their sidearms, but the number of shots fired by each man remains unclear. Zuyev successfully wounded the mechanic; but Zuyev himself was shot in the right arm.

Taking advantage of his opponent's gunshot wound, however, Zuyev jumped into the MiG-29's cockpit. "But when I tried to start the engines," he said, "it was just a dry bone click."

He tried it again.

"It didn't work."

Panicking, Zuyev, turned the ignition for a third time, whereupon the MiG-29's engines finally sprang to life.

After taking off, Zuyev attempted to strafe the airfield, hoping to destroy the other MiGs on the ground before their pilots could give chase. But this strafing maneuver failed because he had forgotten to remove one of the safety locks from the gun.

Realizing that time was running out, Zuyev set a hasty course to the city of Trabzon, Turkey, some 150 miles south across the Black Sea. Considering the MiG's airspeed, however, the Turkish border would only be a 10-minute flight from Mikha Tskhakaya.

As Zuyev vectored over the Black Sea, another MiG-29 from the 176th Regiment (piloted by the Regimental Commander), scrambled to intercept him, but the pursuing MiG couldn't get within firing range before Zuyev entered Turkish airspace. Almost simultaneously, Soviet air defense batteries were alerted of his flight path, but none could find him as he was flying at altitudes well below the radars' envelope.

When he landed at the airport in Trabzon, Turkey, his first words were: *"Finally, I am an American!"*

Turkish authorities treated his wounds, and handed him over to the Americans when he requested asylum.

However, the stolen MiG-29 would not be the same intelligence bonanza that the US had seen with Belenko's MiG-25 in Japan.

Turkey had been a NATO member since 1952; but they shared a land border with the Soviet Union, and the Turkish government was in *no* position to stonewall its neighbor to the east. Thus, when the Soviets demanded the return of their airplane, the Turks happily obliged. The following day, a Soviet recovery team arrived in Trabzon to fly the stolen MiG back to Georgia. Meanwhile, the seven comrades whom Zuyev had tranquilized with his sedative cake were hospitalized following the incident. Fortunately, all seven made a full recovery.

Meanwhile, Zuyev arrived in Virginia, where he was debriefed by a Marine Corps pilot, Lieutenant Colonel Harry Spies. Spies, although bewildered by the account of Zuyev's escape, nonetheless found his story to be credible.

As Zuyev began his new life in America, he began working as a consultant for the CIA and Department of Defense, providing technical data on the MiG-29 and discussing air-to-air tactics used by frontline Soviet forces. This information helped American forces during Operation Desert Storm because the Soviets had trained the Iraqi Air Force and supplied them with nearly forty MiG-29s during the 1980s.

Zuyev eventually settled in San Diego, California where he opened his own aviation consulting firm. He later wrote about his experiences as a Soviet dissident in the bestselling book, *Fulcrum: A Top Gun Pilot's Escape from the Soviet Empire*. Over the next decade, Zuyev led a relatively quiet life – participating in air shows and making occasional public speaking appearances. Sadly, Zuyev was killed on June 10, 2001 along with fellow aviator, Jerry "Mike" Warren, in a crash near Bellingham, Washington, when their Yakovlev Yak-52 failed to recover from an accelerated stall. He was 39 years old.

Zuyev remains one of the most prominent

defectors of the Cold War. But his legacy remains shrouded by contradictory reports. Western media outlets have often praised him for his bravery, resourcefulness, and moral conviction. Eastern Bloc sources, however, castigate him as a reckless traitor with a fragile and over-inflated ego. Like Viktor Belenko before him, Zuyev, too, was beset by marital problems. At the age of 25, he had married the daughter of a high-ranking Soviet Air Forces general. Their union was not a happy one; but unlike Belenko, Mrs. Zuyev had expressed no overtures of divorce or parental alienation. And Mrs. Zuyev was, in fact, pregnant when her husband made the tranquilized cake for his comrades at the airbase. After her husband's defection, Mrs. Zuyev gave birth to their son, Dmitry, who tragically never met his father.

Nearly all sources have confirmed the validity of Zuyev's cake-baking story, but these same sources vary when discussing (a) his service record; (b) his motivations for defecting; (c) the sequence of events at the Georgian airfield; and (d) the details of his "duel" with the mechanic.

Still, there can be little question that Alexander Zuyev made a daring escape. At the twilight of the Cold War, he brought the elusive MiG-29 closer to Western authorities than at any other time before the fall of the Iron Curtain.

Chapter 3:
The Baghdad Express

Saddam Hussein, the President of Iraq, rose to power in 1968 following the Ba'ath Party revolution. As he ascended to the presidency, however, Saddam ruled Iraq with a brand of brutality reminiscent of Hitler and Stalin. Consolidating his power into a dictatorship, he seemed poised for a long, prosperous rule...until his fortunes changed in the wake of the Iranian Revolution.

Although technically allies, the Iraqis and Iranians had never fully trusted one another. Border disputes between the two countries had been ongoing for years. And since the death of Egyptian President Gamal Nasser in 1970, Iraq and Iran had been jockeying to become the dominant force in the Middle East. After the Shah's downfall, however, Saddam took warning of the Ayatollah Khomeini's rhetoric. Saddam Hussein, and many within his Ba'ath Party government, were Sunni Muslims. The majority of Iraq's citizens, however, were Shiite Muslims – just like Khomeini and his disciples. Fearing that the Ayatollah's rhetoric would galvanize Iraq's Shiite majority, the "Butcher of Baghdad" launched a preemptive invasion of Iran on September 22, 1980.

Simultaneously, Saddam had hoped to take advantage of the instability following the Shah's exile and seize key oil fields along the Iranian border. For the next eight years, the ensuing Iran-Iraq War cost thousands of lives and ended in a bloody stalemate.

In the years leading up to the Iran-Iraq War, however, Saddam Hussein had garnered close ties with the Soviet Union. In 1971, he signed an official Treaty of Friendship with the USSR, granting him access to the latest in Soviet fighter aircraft. Throughout the 1970s, Saddam steadily grew the size of his air force, equipping it with modern fighters such as the MiG-23, Su-22, MiG-25, and eventually the MiG-29.

Between 1987 and 1989, the Iraqi Air Force received at least thirty-seven MiG-29 9.12B (single-seat) and MiG-29 UB (two-seat trainer) variants. These export variants, however, were the "bottom-of-the-barrel" MiGs, intended for sale to the lowest-ranking allies beyond the Warsaw Pact. In fact, the "9.12B" suffix was, officially, the *lowest-designated base-model export* for a non-Warsaw Pact country. It featured a simplified NO19EB radar coupled with an OEPrNK-29E2 targeting/navigation system and was armed with R-60M missiles. Aside from the Iraqi Air Force, the Soviets had also given these downgraded variants to the Indian, Syrian, and Yugoslav Air Forces.

Iraq received twenty-five of these MiGs in 1987, towards the end of their war with Iran. However, it remains unclear whether any of these MiG-29s ultimately served during the Iran-Iraq conflict. There are a few anecdotal reports of Iraqi Fulcrums engaging Iranian F-14s, but neither side has produced any official documentation confirming a MiG-29 standoff. Moreover, both the Iraqis and Iranians were notorious for exaggerating their combat claims.

Still, by 1988, the MiG-29 was well in service with the Iraqi Air Force. Much like their Syrian counterparts, many of the Iraqi Fulcrums were painted in a desert camouflage pattern. In keeping with the plane's intended design and function, the Iraqi Air Force used the MiG-29 as a Defensive Counter Air (DCA) fighter, operating from bases at Al-Habbaniya and Tammuz.

Meanwhile, Saddam and the Ayatollah Khomeini had grown increasingly tired of the war. After eight years of conflict, neither side had made any significant gains against the other. Thus, the UN negotiated a peace settlement; and, under the banner of Resolution 598, Iran and Iraq accepted the terms of the ceasefire.

The Iran-Iraq War formally ended on August 20, 1988.

Although the MiG-29 arrived too late to make any impact against the Ayatollah's air forces, the Fulcrum would have its "baptism by fire" in the skies over Babylon.

Aside from the untold cost in human suffering, the Iran-Iraq War left Saddam Hussein straddled with a multi-billion-dollar debt – most of which had been financed by Kuwait. Prior to the war, Iraq had almost no foreign debt and more than $35 billion in cash reserves. By 1989, however, Iraq had spent nearly $60 billion in arms purchases alone. But rather than pay his debt to the Kuwaiti government, the "Butcher of Baghdad" simply invaded his neighbor to the south.

To justify the invasion, Saddam reignited the long-standing border dispute between the two countries. He also made false allegations that the Kuwaitis had been slant-drilling Iraqi oil and claimed they were deliberately trying to keep oil prices low by producing beyond OPEC's quotas. Kuwait held ten percent of the world's oil reserves and generated 97 billion barrels of crude each year. Thus, Saddam reasoned that if he could not repay his debt, he would simply annex the tiny emirate and take over its petroleum industry.

Thus, on the morning of August 2, 1990, more than 100,000 Iraqi troops and several hundred Iraqi tanks stormed across the border, the spearhead of an eighty-mile blitzkrieg into Kuwait City. Encountering only piecemeal resistance, Iraqi tanks thundered into the heart of the Kuwaiti capital.

The invasion drew fierce condemnation from the international community, prompting the UN to demand Saddam's withdrawal. Undeterred by the rhetoric, the Iraqi dictator massed his forces along the Saudi Arabian border and dared the world to stop him. He was certain that his army – the fourth-largest in the world and equipped with the latest in Soviet weapons – would make short order of any rescue force that came to liberate Kuwait.

Economic and military sanctions soon followed while President George HW Bush authorized the first US deployments to the region. Dubbed "Operation Desert Shield," it was a preemptive deterrent against Saddam Hussein lest he try to invade the Kingdom of Saud.

But for as tough as the Iraqi Army appeared on paper, Saddam's air force was primitive by NATO standards. In fact, much of the Iraqi Air Force's vitality had been eroded by the Iran-Iraq War. Still, it was the largest air force in the Middle East, with some 934 combat-capable aircraft in its arsenal, including the MiG-29. Theoretically, the Iraqi Air Force should have been "battle-hardened" by their conflict with Iran, but Saddam's postwar purges had stripped the Air Force of its best leaders. In the wake of this brutal crackdown, training and readiness within the Iraqi Air Force quickly ground to a halt.

In November 1990, as coalition forces poured into Saudi Arabia, the UN passed Resolution 678. The resolution, for what it was worth, gave Saddam Hussein a deadline of January 15 to withdraw his forces, or face military action.

Still, the Iraqi dictator showed no signs of backing down.

Thus, it came as no surprise when, on the morning

of January 15, 1991, coalition forces awoke to the news that Saddam Hussein had reached his deadline – and had made no effort to withdraw from Kuwait. The next day, President Bush announced the start of the military campaign to eject the Iraqis from the war-torn emirate.

Operation Desert Shield had just become Operation Desert Storm.

On January 17, the first wave of the coalition's air campaign destroyed Iraqi radar sites near the Saudi border. For the next five weeks, coalition air forces pounded away at key targets within Iraq and Kuwait.

Saddam, however, had not been idle during the buildup to Desert Storm. He had prepared his fighter and interceptor squadrons at air bases including Al-Taqqadum, Al-Asad, Mudaysis, H-2, and H-3. Unlike the Iraqi Army, whose frontline formations were manned mostly by conscripts, the Iraqi Air Force had the best-trained personnel. Saddam confidently instructed his airmen to destroy the "infidels" who were about the descend onto Iraq.

The plan for the first night of the Allied air campaign was a "multi-pronged attack across Iraq's air defense network." F-117 Stealth Fighters would hit key targets over Baghdad while F-15E Strike Eagles would fly lower-altitude missions against the H-2 and H-3 airfields. Simultaneously, EF-111 electronic warfare planes would fly in low, jamming Iraqi surveillance and air defense radars to facilitate the F-117's ingress and egress. Contrary to popular belief, "stealth" capability did not mean that the F-117s were completely invisible to enemy radar. Rather, it meant they had a "low observable" radar signature.

Once these F-117s, F-15Es, and the accompanying cruise missiles found their mark, however, the Allied coalition no longer had the element of surprise. As expected, the Iraqis scrambled their MiG-29s to interdict the Allied aircraft.

As the initial wave of F-117s and Strike Eagles egressed from Iraqi airspace, eight F-15s from the 58th Fighter Squadron (USAF) swept north to interdict any enemy fighters responding to the initial bombardment. Taking off from Tabuk Air Base in Saudi Arabia, these F-15s were divided into two, four-ship flights codenamed PENNZOIL and CITGO.

This inaugural mission into Iraq would mark the first encounter between an American F-15 and a MiG-29.

For the F-15s, however, their first stop was a rendezvous with the KC-135 Stratotankers, all of which were flying at distances beyond the range of Iraqi air defense radars. In the middle of their refuel, however, the PENNZOIL flight leader received an urgent call from the nearby AWACS. Enemy fighters had been detected scrambling from the southern airfields, on a vector that would put them directly in path of the egressing Strike Eagles. With these new mission parameters, PENNZOIL and CITGO hastily merged, pushing forward into a linear "Wall of Eagles" formation.

One of the pilots manning this "Wall of Eagles" was Captain Jon "JB" Kelk. As Kelk recalled: "The big unknown was the capability and response of the Iraqi Air Force. If you just look at it on paper, it was a formidable force in terms of total numbers of fighter aircraft." All told, Kelk and his comrades expected to encounter a few MiG-25s and MiG-29s. By 1991, most of the Western world knew that the MiG-25 was fast, but substandard as a dogfighter.

The MiG-29, however, was a different story. Indeed, the MiG-29 was purportedly capable of standing toe-to-toe against *any* fighter in NATO's inventory. But beyond these hearsay reports and spotty intel, Western pilots knew virtually nothing about the Fulcrum. As Kelk's fellow pilot, Cesar "Rico" Rodriguez said: "We thought the MiG-29 was the biggest, baddest kid on the block." But come what may, Kelk and his comrades were prepared to take on the Iraqi airmen.

Pushing forward into the linear formation, JB Kelk and his PENNZOIL wingmen flew northward, scanning their radars and listening intently to the radio traffic. "The radios got very busy on the strike frequency," he recalled.

Suddenly, Kelk's radar showed the blip of an incoming aircraft.

A bandit? - he wondered.

At a distance of 40 miles north, altitude 7,000 feet,

its relative position indicated that it was likely an enemy bandit. "I tried to get the AWACS to confirm...but had no luck. I was still very concerned about frat [fratricide], but even without the AWACS's help, I was confident that this was a bad guy. He was heading south towards me and climbing."

Indeed, by the time Kelk readied his first missile, the bandit was at 17,000 feet. "Once I was in parameters," he said, "I fired a single AIM-7." As soon as Kelk depressed the trigger, he closed his eyes, not wanting the white flash of the missile to rob him of his night vision.

After deploying his missile, however, Kelk was unsure of whether the AIM-7 had met its target. "I look out the front and I see a purple-blueish light on the horizon" – not the telltale orange fireball he had been expecting. "After about three to five seconds, it fades."

Kelk was bewildered.

Did he kill the bandit?

Was this purple-blue ember the skewed sight of an explosion?

Had the enemy bandit been dropping flares?
He did a radar search of the immediate area and found nothing. Whoever this bandit may have been, he had now disappeared from radar view.

Back at the Tabuk airbase, still unsure of his aerial victory, Kelk reported it as a "probable" kill. A few hours later, however, the AWACS that had been aloft with PENNZOIL sent its report to the 58th Fighter

Squadron. Not only had Kelk destroyed the bandit, but the kill had been confirmed as a MiG-29.

Jon "JB" Kelk made history that night.

He had achieved the first aerial kill of Desert Storm and the first-ever combat kill of a MiG-29.

Later that same day, one of Kelk's fellow pilots in the 58th Fighter Squadron, Captain Charles "Sly" Magill, became the first Marine aviator to earn a confirmed kill since Vietnam.

Magill was an "exchange pilot" from the Marine Corps who had been temporarily assigned to the 58th Fighter Squadron in the fall of 1989. By the time he arrived at the 58th, Magill had already logged more than 1,000 hours in the F/A-18 Hornet. Settling into his new "broadening assignment," however, Magill had become quite adept at piloting the F-15.

Now on the morning of January 17, he would be leading a flight of F-15s into combat, covering a strike force of forty F-16s. Flying into Iraq, these F-16s would be accompanied by F-4 Wild Weasels and EF-111 Ravens: the former to suppress enemy air defenses; the latter to disrupt the Iraqis' radar network. According to the plan, Magill's eight-ship formation would sweep 25-30 nautical miles ahead of the main body "strike package" and destroy any loitering Iraqi fighters.

For this new sortie, the 58th Fighter Squadron's formations once again featured the callsigns CITGO and PENNZOIL, this time manned by different pilots.

Magill's CITGO four-ship would occupy the western sector, sweeping to the northwest side of Baghdad near Al Asad Air Base.

Launching with the second iteration of the CITGO flight, Magill went airborne alongside PENNZOIL to his right flank. Almost immediately, the nearby AWACS alerted them of two incoming bogeys.

Both were identified as MiG-29s.

As CITGO and PENNZOIL, continued pushing northward, it appeared that the MiGs were travelling north-south at a lower altitude. Magill could see the Fulcrums on his radar, and soon confirmed that they were 30 miles out. Turning his formation onto a 30-degree bearing, Magill hoped to meet the Fulcrums head on. However, as they closed within 22 miles, he was shocked when the radar showed him the bandits "turning cold" – meaning that the MiG-29s had now turned *away* from the F-15s and were now heading north.

But Magill wasn't going to let these MiGs get away. Accelerating to full afterburner, he and his wingmen closed within 16 miles when the bandits suddenly turned back to meet the F-15s head on – accelerating to 600 knots. At 14 miles out, one of Magill's wingmen, Rhory Draeger, got missile lock on one of the MiGs—firing a single AIM-7.

"Fox 1!" Draeger cried.

The Sparrow missile found its mark on one of the bandits, striking the MiG-29 right through its canopy. Magill, meanwhile, set his own lock onto the second

Fulcrum, which was closing in at about 1,250 knots.

From a distance of 12 miles, Magill fired his own AIM-7. However, the problematic folding-fin rocket malfunctioned and went barreling towards the ground. Not impressed by the AIM-7's wild trajectory, Magill lined up another shot and fired his second Sparrow of the day.

But, as it turned out, Magill's first missile (with its seemingly errant trajectory), corrected itself and impacted the MiG-29 just below its right wing. As the Fulcrum began teetering downward, Magill's second missile went right through the center of the MiG-29's fuselage. Egressing from the area, Magill radioed in to the nearby AWACS: "Splash 2! Splash 2!" – indicating that both MiGs had been destroyed.

Years after the war, Chuck Magill discovered why the Iraqi Fulcrums had suddenly turned back towards the F-15s after previously withdrawing northward. As it turned out, the MiGs had been trying to intercept a flight of F-14s that were egressing from western Iraq. The prowling MiGs, however, had run right into Magill and his comrades in the CITGO flight.

All the while, the Tomcats had been about nine miles ahead of the MiG-29s. However, when the F-14s' flight leader finally saw the trailing Fulcrums, and turned his formation to meet the incoming bandits, he was startled to see Magill's F-15s fly right over him and destroy the MiGs on their own. Magill later guffawed: "We never even saw those Tomcats under us!"

On January 19, 1991, Captains Cesar "Rico" Rodriguez and Craig "Mole" Underhill – also from the 58th Fighter Squadron – each scored an additional kill of a MiG-29. As the flight lead of a four-ship formation, Rodriguez and his wingmen received a pop-up alert from the nearby AWACS. "Our new mission," said Rodriguez, "was a pop-up tasking to strike a newly-discovered weapons storage area southwest of Baghdad. The plan was to put 20-30 F-16s onto this target."

According to the AWACS, four F-15s would fly in front of the strike formation, conducting a classic pre-strike fighter sweep, while Rico and Mole would fly behind the formation as a "post-strike sweep" to conduct battle damage assessments and engage any lingering MiGs.

"So, as we arrived on the scene," he said, "everything was looking good." But suddenly, the AWACS broke in with an urgent call:

"CITGO 25, pop-up contacts, 40 north of the target area."

Rodriguez had been expecting this.

The AWACS call indicated that two unknown aircraft, likely enemy, had been detected approximately forty miles north. "Sure enough, at about 60 miles from us, we picked up two unidentified targets," said Rodriguez. "We were up at 30,000 feet, and we proceeded northeast as we were watching the [F-16s] attacking."

As Rodriguez and Underhill vectored to the northeast, both pilots saw the F-16 formation heading south—thus confirming that the F-16s had completed their strike mission.

Just then, however, the trailing F-16 pilot radioed that he had fallen under radar lock from one of the two unidentified aircraft coming from the north. At this point, Rodriguez wasn't sure what kind of aircraft was pursuing the F-16s, but to close within a 35-mile distance that quickly, the offending aircraft had to be a fighter.

The F-16s, meanwhile, weren't fazed or even threatened by the lock-on, and they continued egressing southbound out of the target area—just in time for Rodriguez and Underhill to intercept.

But once the F-15s closed within 18 nautical miles, both Iraqi planes suddenly reversed their course and retreated northward.

After confirming from the AWACS that there were no other friendlies in the area, Rodriguez and Underhill gave chase to the fleeing bandits. Around this time, both men confirmed that the elusive jets were, in fact, MiG-29s.

The MiGs, for their part, desperately tried to shake off the incoming F-15s. They dove to a lower altitude and were pushing the Fulcrum's engines to nearly as fast as they could run.

While closing in on the Iraqi MiGs, Rodriguez and Underhill realized that they were quickly approaching Baghdad airspace. Within moments, their radar warning systems began to sound the alert of incoming SAMs. Both men were about to clear the airspace when the AWACS interrupted, saying that the bandits were now *13 miles out* on a heading of 330 degrees.

Almost instinctively, Rodriguez slammed his throttle to full afterburner, pulling 9g's into the turn to bring his F-15 onto a 330 heading while he jettisoned his external fuel tanks. Craig Underhill followed suit and, within moments, both F-15s gained visual contact of the lead MiG-29.

"At this point, that MiG-29 locks me up," – the radar system indicated as much. "I execute a defensive maneuver" – dispensing flares to confuse the MiG-29's radar. Meanwhile, Underhill sprang into action against the leading Fulcrum. Locking onto the leader, Underhill fired an AIM-7.

"Fox 1!" he cried.

Watching the missile glide off Underhill's wing, Rodriguez watched intently as the AIM-7 impacted squarely on its nose, sending the MiG-29 into a brilliant flash of molten steel. Elated, Rodriguez called "Splash one!" to the AWACS, confirming his wingman had just destroyed an enemy bandit.

But there was no time to celebrate.

Mere moments after the AWACS had confirmed Rodriguez's report, they issued another warning:

"Second bandit, three miles in trail!"

By this point, Rodriguez and Underhill were both flying at a low altitude, about two and a half miles apart, and the incoming bandit was closing fast to their six o'clock position. Both F-15s executed a hard-right turn to meet the incoming bandit head-on.

But identifying this Fulcrum would be more problematic.

Underhill locked onto the bandit with his radar but couldn't positively identify it as "friendly" or "hostile." The ever-present risk of fratricide pressured every pilot to err on the side of caution if he couldn't identify the bogey from beyond visual range.

Today, as neither Rodriguez nor Underhill could identify the bandit using their onboard instruments, both men continued flying forward to identify it *visually*.

As the bandit approached, Rodriguez knew it would be difficult to identify the MiG-29 solely based on its front form and silhouette. Indeed, from greater distances, the MiG-29's silhouette could easily be mistaken for an F-15. Thus, Rodriguez decided to "bracket" the flight formation, which allowed him to set up a visual ID by flying underneath the bandit while flying directly into its flight path. Using this method, Rodriguez flew under the bandit's left wing by nearly 100 feet – close enough to detect the unmistakable markings and paint schemes of the Iraqi Air Force.

As Rodriguez pulled behind the MiG-29, its pilot slowly began turning left – demonstrably unaware that an American F-15 was closing in behind him. At some point, however, the Iraqi pilot was alerted to

Rodriguez's presence, because the MiG-29 suddenly accelerated, vectoring into a defensive turn. Closing in to nearly 3,000 feet, Rodriguez elevated the nose of his aircraft to align an AIM-7 missile shot.

Meanwhile, the MiG-29 pilot, perhaps in a state of panic, tried to shake Rodriguez by performing a "Split-S" maneuver. According to Rodriguez, the Split-S was a defensive action wherein an enemy plane rolls upside down, pulls downward, and reverses course – usually as a means to disengage from combat.

However, while trying to evade Rodriguez, the Iraqi pilot inadvertently plummeted himself into the ground. When the MiG-29 hit the desert floor, Rodriguez noted that the plane was nearly perpendicular to the ground. Even after the MiG made impact, Rodriguez could see its rear stabilizers sticking up from the ground...with the afterburners still running.

Meanwhile, on the Iraqi side of the border, Saddam Hussein was already against the ropes. At airbases across Iraq, much of the Iraqi Air Force was being destroyed before it even got off the ground.

But the MiG-29s, for their part, had not been idle. On January 19, 1991, Iraqi Air Force Captain Jameel Sayhood claimed to have shot down an RAF Tornado with his MiG-29. According to Sayhood, the Tornado was making a low-level bombing attack on the Ar-Rutbah radar site. He also claimed to have taken the

Tornado by surprise, downing the aircraft with his R-60MK heat-seeking missile.

However, the details of Sayhood's engagement are anecdotal at best. Indeed, Western analysts have disproven his claims based on multiple corroborated records. For example, the flight data for January 19 shows the only Tornado that Captain Sayhood could have been referring to was "ZA467," which was shot down three days later, January 22, and confirmed as a surface-to-air missile (SAM) loss. Another RAF Tornado (ZA396) was indeed shot down on January 19, but it belonged to a different squadron, and it was travelling along a different flight path. That Tornado, too, was confirmed to be a SAM kill.

Moreover, there's evidence to suggest that Jameel Sayhood was the Iraqi pilot who crashed his MiG-29 while trying to evade Cesar Rodriguez during the aforesaid F-15 duel of January 19. Thus, it'd be unlikely (although not impossible) that Sayhood could have ejected in time to survive the crash from his errant Split-S maneuver.

If Captain Sayhood's tale is fictional, it would be one among several exaggerated and/or fabricated accounts perpetrated by Iraqi pilots. For example, one Iraqi aviator claimed to have shot down an F-15, but no F-15s were downed by enemy fighters during that conflict. Other Iraqi pilots claimed shooting down Allied planes that were later confirmed to have crashed.

Surprisingly, by the third week of the air campaign, several Iraqi pilots had begun flying their jets into *Iran*, hoping to use the shield of Iranian neutrality as a means to wait out the conflict. By mid-February 1991, the air-to-air engagements of Desert Storm had all but ceased.

However, as the air war dragged on into its second month, coalition ground forces prepared for their initial assault into Iraq. On the evening of February 21, 1991, President Bush issued his final ultimatum: Saddam Hussein had twenty-four hours to withdraw from Kuwait or face destruction at the hands of coalition forces. Despite this ultimatum, however, the Iraqi dictator dug in his heels and told his troops to prepare for combat. From these developments, one thing was certain: the anticipated "Mother of All Battles" was coming to pass.

On February 24, coalition ground forces began their initial push into Iraq. Although defense analysts and the normal variety of "experts" had anticipated several thousand Allied casualties, they were surprised to see Iraqi units (even the vaunted Republican Guard) being outclassed, outgunned, and outmaneuvered by American forces.

Finally, on February 28, President Bush announced a cease-fire to the ground war. Barely 100 hours after the start of the Allied invasion, the Iraqis were in full retreat and Saddam was suing for peace. The "Mother of All Battles" had ended, and it was the Iraqi Army that had been destroyed. In their disastrous retreat, the Iraqis had fled Kuwait, leaving a devastated country in their wake. On March 3, 1991, General H. Norman Schwarzkopf, commander of UN Forces, met with several Iraqi generals in Safwan to discuss the terms of surrender.

The Iraqi Air Force, meanwhile, lay in shambles. They had lost more than 250 aircraft while confirming only four air-to-air victories of their own. Many of the Iraqi pilots who fled to Iran were later repatriated, but their jets were impounded by the Ayatollah's government. Most of these Iraqi airframes were impressed into Iranian service, with the Ayatollah claiming them as "reparations" for the Iran-Iraq War. The Iranians did, however, eventually return a handful of Sukhoi Su-25 attack planes.

As for the Iraqi MiG-29s, there was no escaping the conclusion that the Fulcrum had been outclassed by the American F-15C. Assessing the MiG-29's performance during the Persian Gulf War, however, must be viewed against the comparative skill of the pilots involved. Most experts agree that the Iraqi Air Force suffered from a three-tiered crisis of logistics, maintenance, and culture.

Indeed, the Iran-Iraq War had eroded Saddam's air force. By 1990, the Iraqi Air Force was already struggling to maintain its fleet of MiG-29s. Eight years of war had rendered their supply system short on parts and technical expertise (the latter being a function of Saddam's perennial "purges" throughout the ranks). To boot, the Iraqi Air Force simply didn't have a well-trained cadre of experienced pilots. The complexity of the MiG-29 demanded highly-skilled aviators who could leverage the aircraft in a dogfight. Iraqi Air Force leaders, however, never developed or maintained an adequate training program for the MiG-29, leaving their pilots ill-prepared for air-to-air combat during the Gulf War.

Then, too, Arab pilots (and their native air forces) had never been known for their flying prowess. In the Mediterranean, for example, the Syrian and Egyptian Air Forces had been routinely trounced by the IDF. In Libya, meanwhile, Muammar al-Gaddafi's air force had been bested by US Naval air squadrons on numerous occasions.

Russian air historian Yefim Gordon later suggested that, given the sad state of affairs within the Iraqi Air Force, the outcome of the Gulf War was a foregone conclusion – "considering that in modern warfare, the outcome of a battle depends on pilot skill and adequate information support as much as on the aircraft's capabilities – and the Iraqi pilots were definitely lacking in the first and second."

By the end of the Gulf War, the Iraqi Air Force had only *twelve* functional MiG-29s remaining in service. Ironically, it was just as well that the Iraqis had lost most of their MiG-29s. Apparently, Iraqi Air Force leaders had never fully warmed up to the Fulcrum. Even before the invasion of Kuwait, Saddam cancelled his existing orders of the MiG-29, opting instead for a new order of Su-27s. However, because of the 1990 UN Arms Embargo, his Sukhoi deliveries never arrived.

The last Iraqi MiG-29s were retired in 1995.

Jon "JB" Kelk. On January 17, 1991, Kelk made history by achieving the first aerial kill of Desert Storm and the first-ever combat kill of a MiG-29. *US Air Force*

Photo Section

Alexander Zuyev in an undated picture following his 1989 defection from the Soviet Union. A dedicated MiG-29 pilot, Zuyev escaped from the USSR, stealing his own MiG-29 from an airbase in Soviet Georgia and flying it to Turkey, seeking asylum in the United States. Zuyev passed away in 2001. *CC-BY-SA-4.0/OHFMITRY*

An East German MiG-29 at Preschen Airfield, 1990. *Rob Schleiffert*

Czechoslovakian MiG-29s at the International Air Tattoo, RAF Boscombe Down (UK), June 1992. By the following year, the Czechoslovak Air Force would cease to exist. Partitioning into Slovakia and the Czech Republic, the new respective air forces each retained a portion of the legacy Czechoslovak MiG-29s. *Andrew Thomas*

Hungarian Air Force MiG-29 at the Royal International Air Tattoo, July 2008. *Tim Felce*

Polish Air Force MiG-29 at the ILA Berlin Air Show, 2016. This Fulcrum bears the subdued commemorative roundel of the famed "Kościuszko Squadron" – the Polish 7th Air Escadrille, founded by American aviator Merian Cooper in 1919. *Julian Herzog*

The charred remains of an Iraqi MiG-29, destroyed by coalition forces during Operation Desert Storm, 1991. *US Air Force*

Cesar "Rico" Rodriguez, American F-15 pilot and famed "MiG Killer." Assigned to the 58th Fighter Squadron, he successfully downed an Iraqi MiG-29 and MiG-23 during Operation Desert Storm, 1991. Eight years later, he would shoot down another MiG-29 during the NATO air campaigns over the former Yugoslavia. *US Air Force*

Specialist Daniel Jackson, a US Army combat cameraman, inspects the wreckage of a downed Iraqi MiG-29, February 1991. *US Department of Defense*

A MiG-29 from the Islamic Republic of Iran Air Force (IRIAF) prepares to take off from the Shahid Babaei Air Base near Isfahan. Like their Iraqi neighbors, the IRIAF acquired the MiG-29 in the late 1980s, towards the end of the Iran-Iraq War. *Abbas Pustinduz, Mehr News Agency*.

An Indian Air Force (IAF) MiG-29 at Yelahanka Air Force Station in Bangalore, India. The IAF has operated variants of the MiG-29 since 1987 and used it to great effect during the Kargil War of 1999. *Aeroprints*

A US Army documentation team surveys the wreckage of a Serb/Yugoslav MiG-29 near Ugljevik in Bosnia & Herzegovina, on March 27, 1999. It was one of a handful of MiG-29s shot down by NATO forces during Operation Allied Force. *US Department of Defense*

The wreckage of another Serb/Yugoslav MiG-29 following Operation Allied Force, March 1999. The Fulcrum pictured here was downed by the US 493d Fighter Squadron. *US Air Force*

Captain 1st Class Zoran Radosavljevic, Serbian MiG-29 pilot pictured here in 1997. After being shot down by the US 493d Fighter Squadron on March 26, 1999, his plane crashed south of Bijeljina. Although he ejected from the stricken plane, Radosavljevic did not survive the ejection. His remains were recovered by local villagers who facilitated his extradition back into Serbian hands. Radosavljevic was buried three days later at Lešće Cemetery in Belgrade. *Serbian Ministry of Defense*

American F-15 pilot Jeff Hwang (pictured here at the rank of Colonel) displays the "victory markings" on the side of his aircraft, with each star designating an aerial victory. The double stars commemorate the two Serbian MiG-29s that Hwang shot down in March 1999 while assigned to the 493d Fighter Squadron. One of the ill-fated MiGs was piloted by Captain 1st Class Zoran Radosavljevic, pictured on the preceding page. *US Air Force*

An Eritrean MiG-29. Following its independence from Ethiopia, the new Eritrean Air Force acquired a handful of MiG-29s. Although undermanned, undertrained, and under-resourced, the Eritrean MiGs nonetheless fought valiantly against the Ethiopian Air Force during their 1999 border conflict. *Madote News Service*

A Sudanese Air Force MiG-29SE in flight during the Darfur Insurgency, 2007. Since their arrival in Sudan, the MiG-29s have flown combat missions against JEM guerrillas and the breakaway South Sudanese forces. *Melting Tarmac Images*

A Chadian Air Force MiG-29 at Lviv International Airport in Ukraine, 2014. Chad imported a handful of MiG-29s from Ukraine during the 2010s, but none are known to be flyable as of 2024. *Oleg Volkov*

Bangladeshi MiG-29UB. One of the smallest operators of MiG-29s, the Bangladesh Air Force acquired its first Fulcrums in 1999. *Bangladesh Air Force*

Sporting a black color scheme, a Russian MiG-29K performs an aerial demonstration at the International Aviation and Space Show (MAKS), 2003. *Leonid Faerberg*

A brightly-colored MiG-29UB during an aerial demonstration of the Russian Air Force's "Swifts" aerobatic team. Formed in April 1991, the Swifts perform at airshows worldwide in a manner similar to the USAF Thunderbirds and the US Navy's Blue Angels. *Aleksandr Markin*

Top, Center, and Bottom: Ukrainian MiG-29s from the 40th Tactical Aviation Brigade prepare for action on the eve of the Russian invasion; February 21, 2022. *Ukrainian Defense Ministry*

Lieutenant Colonel Vyacheslav Yerko (posthumously promoted to full Colonel), a MiG-29 pilot from the 40th Tactical Aviation Brigade who perished on the first day of the air war, February 24, 2022. As the lead pilot of a two-ship formation, Yerko successfully downed a Russian Su-25 "Frogfoot" attack plane and two Russian Mi-24 helicopters, before being shot down by an enemy Su-35 Flanker, southwest of Kyiv. *Ukrainian Defense Ministry*

Top and Bottom: Artistic renditions of the "Ghost of Kyiv." During the first few days of the air campaign, stories began to circulate that a nameless Ukrainian MiG-29 pilot, known only as the "Ghost of Kyiv," was blasting Russian planes from the sky with impunity. Initial reports claimed that the Ghost had shot down *six* enemy planes within 24 hours, thereby earning him the title "Ace in a Day." The Ghost of Kyiv was later proven to be a hoax, but the story created a media firestorm during the opening months of the Russia-Ukraine War. *Top photo by Andrii Dankovych; Bottom photo by Vischun via DeviantArt.*

(Top) Captain Andrii Pilshchykov, Ukrainian MiG-29 pilot in a photo dated April 25, 2019 near Kyiv. On February 25, 2022, Petro Poroshenko tweeted this photo, claiming it to be the Ghost of Kyiv, which then became viral. (Bottom) Pilshchykov, of course, was not the Ghost of Kyiv, but was nevertheless a valiant pilot who flew more than 100 combat missions over Ukraine before his death during a training accident in August 2023. *Ukrainian Defense Ministry*

Two MiG-29s from the 40th Tactical Aviation Brigade, April 2023. The Fulcrum in the foreground is sporting the customized black color scheme and the emblem of the "Ghost of Kyiv." Although the Ghost was nothing more than a legend, many fighter units within Ukraine's Air Force proudly wore the Ghost roundel as a symbol of rugged defiance in the face of the Russian invasion. *Ukrainian Defense Ministry*

The charred and scattered remains of Lt. Dmytro Shkarevsky's MiG-29. Shkarevsky safely ejected from the plane, but he was unsure of who (or what) had shot him down. He had just engaged a Russian "kamikaze drone" when his own plane began to break apart. As there were no Russian manned aircraft in the area, Shkarevsky blamed his unit's haphazard maintenance program as the reason for his plane's sudden demise. *Ukrainian Defense Ministry*

The MiG-29, "White 12," piloted by Captain Vladyslav Saveliev of the 114th Tactical Aviation Brigade. Saveliev was shot down on June 2, 2023 while flying over the Donetsk region. When the Russia-Ukraine War erupted in February 2022, Saveliev was enrolled in a US-sponsored training program at Columbus Air Force Base, studying advanced fighter tactics and maneuvers. Although eager to return to his home country, Saveliev had to wait an additional twelve months before completing his coursework. He returned to Ukraine in March 2023, nearly three months to the day before his tragic demise. *Andrii Murskyi - Ukrainian Defense Ministry*

Vladyslav Saveliev's official portrait from the 114th Tactical Aviation Brigade. *Ukrainian Ministry of Defense*

The cockpit of this Ukrainian MiG-29 features the onboard adaptive fire control system (consisting primarily of a touchscreen tablet) to accommodate firing the American-made AGM-88 HARM guided missile. In a strange turn of events, NATO began supplying the Ukrainian Air Force with Western-built precision-guided munitions, including the AGM-88 and JDAMs. However, because the MiG-29 wasn't designed to accommodate any Western-style ammunition, Ukrainian engineers and Western contractors developed an adaptive fire control system, wherein the MiG pilot could employ the NATO-spec missiles via commercial iPads and other touchscreen tablets.
Ukrainian Ministry of Defense

Chapter 4: Balkanized

In the aftermath of Desert Storm, the Socialist Federal Republic of Yugoslavia began to fracture along ethnic and religious lines. Yugoslavia, a "non-aligned" Communist state, had been the dominion of President Josip Broz Tito. Following Tito's death and the collapse of Communism in Eastern Europe, however, the diverse ethnic and religious groups within Yugoslavia (including Serbs, Croats, Bosnians, et al) began suing for independence. These independence movements, however, soon devolved into a civil war—whereupon NATO eventually intervened.

Like their counterparts in the former Soviet Union, the Yugoslav Air Force saw much of its inventory absorbed into the successor states. Unlike the Soviet Union, however, these former Yugoslavian entities turned their weapons against each other.

At its peak in 1990, the united Yugoslav Air Force had a full contingent of Soviet and domestically-built aircraft, the former of which included the MiG-29. In fact, Yugoslavia became the first foreign, non-Warsaw Pact recipient of the MiG-29 in 1987, when it ordered a squadron-sized complement of fighters from the USSR. However, the Yugoslav government regarded the MiG-29 as a temporary solution until the Bosnian-based Soko aircraft company could build its own fourth-generation fighter. These imported MiG-29s were assigned to the 204th Fighter Regiment at Batajnica Air Base near Belgrade.

When the Yugoslav Wars began, Serbian forces effectively took control of the air assets in Belgrade. With the Serb-dominated federal government seeing itself as the true heir to President Tito's legacy, they began launching airstrikes against Croatian targets in August 1991.

Thus began the second chapter of the MiG-29's combat history.

Throughout the Yugoslavian Civil War, the MiG-29's role was quite limited; it was used primarily for ground attack missions. For instance, on October 7, 1991, two MiG-29s were *purportedly* involved in an air strike on the Banski Dvori palace, then-home to Croatia's President, Franjo Tudjman. Despite being only a few meters from the impact area, Tudjman survived the airstrike. The following year, he permanently relocated his quarters to the modern-day Presidential Palace in Zagreb.

Although the bombing of Banski Dvori was a bold surgical strike, some aviation analysts have doubted whether a MiG-29 was the true culprit. The offending missile was confirmed to be an AGM-65 – an American-made weapon to which the MiG-29 had not been adapted. All told, it is more likely that the missile came from a Yugoslav J-22 attack aircraft, as they had been specially-modified to carry the AGM-65.

Although the MiG-29's role in the Banski Dvori airstrike remains inconclusive, a MiG-29 did, in fact, destroy a handful of Croatian Antonov An-2 liaison planes on the ground at the Osijek-Čepin airfield later that year.

These domestic air attacks, however, quickly ended with the onset of NATO's intervention. Beginning with Operation Sky Monitor in November 1992, followed by Operation Deny Flight in March 1993, NATO aircraft gradually increased their role in the skies over Yugoslavia. By the summer of 1993, US and NATO aircraft could fly fully-armed, close air support missions in support of UN peacekeepers.

But to the pilots themselves, this "peacekeeping" mission felt like more of a sightseeing tour. The heart-pounding combat missions reminiscent of Desert Storm were few and far between. US Air Force pilot Lieutenant Colonel Michael Arnold, summarized the collective frustrations shared by most NATO pilots at the time: "Most of the missions were very benign and very dull," he said. "We enforced the No-Fly Zone, so we would go up and try to engage anyone who was not supposed to be flying, but of course, we never saw anything."

There were some notable exceptions, however – including the shootdown of American F-16 pilot Captain Scott O' Grady by a Serbian SAM battery in 1995. But on August 30 of that year, NATO began a sustained air campaign against renegade Bosnian Serbs, which ultimately hastened the ratification of the Dayton Accords, and the end of the Bosnian War.

But while Bosnia and Croatia celebrated their independence, Serbia and Montenegro continued their political union, operating as the "Federal Republic of Yugoslavia." Consequently, the new Yugoslav Air Force was drawn from remnants of the legacy force that remained under Serbian control. At its peak, the neo-Yugoslavian air defense command had more than a dozen MiG-29s.

Because of the UN Arms Embargo, however, these MiGs began to deteriorate from a lack of spare parts. Meanwhile, as the rump-state government was trying to assert itself as the successor to Tito's Yugoslavia, they soon faced an insurgency in the province of Kosovo. By 1998, this separatist movement had escalated into a full-scale, high-intensity conflict, once again precipitating NATO's intervention.

NATO – already having a foothold in the region from the IFOR/SFOR peacekeeping missions in Bosnia – was now determined to drive the Serbians from Kosovo on the grounds of humanitarian intervention. Thus, on March 24, 1999, the air campaign for Kosovo began under the banner of Operation Allied Force.

According to NATO parameters, Allied Force had three objectives: (1) Ground interdiction to keep Serbian forces out of Kosovo; (2) Provide Close Air Support as needed; and (3) Establish air superiority over the region. The third objective, naturally, implied shooting down any hostile Serbian/Yugoslav aircraft.

On that morning of March 24, however, the Yugoslav air command was prepared for battle. The 204th Fighter Regiment, for example, had redeployed their MiG-29s to critically-positioned airfields. Two Fulcrums went to Niš and Ponikve, with an additional MiG-29 sent to Podgorica. The six remaining airworthy Fulcrums (along with five others that had been grounded for maintenance) remained at the Batajnica Air Base.

Simultaneously, the Yugoslav Air Force redeployed their MiG-21s from the 83d Fighter Regiment, combining them with the MiG-29s into a series of "task forces" at the various airfields. These mixed fighter detachments were consolidated at Niš, Ponikve, and Podgorica Air Bases – each consisting of two MiG-29s and two MiG-21s. A senior Fulcrum pilot commanded each of the task forces, all of whom were older and more experienced than the MiG-21 pilots from the 83rd Regiment.

Meanwhile, on the other side of the Adriatic Sea, US air squadrons in Italy – consisting of F-117 Nighthawks, F/A-18 Hornets, EA-6B Prowlers, F-15C and F-15E Strike Eagles – departed from Aviano Air Base that evening. From other Italian air bases, participating NATO allies launched their own aircraft – including RAF Harriers, German Tornados, Dutch F-16s, and French Mirage fighters.

At 7:30 PM local time, March 24, Yugoslav air

defense radar detected the first wave of incoming aircraft over Albania – an American strike package consisting of EA-6Bs, F-15s, and F-16s. A few minutes later, US and Royal Navy vessels in the Adriatic Sea launched their first cruise missiles against select targets in Yugoslavia. By 8:00 PM, the government of Slobodan Milosevic declared that Yugoslavia was at war "since the aggression had started."

There were two waves of Allied airstrikes on the night of March 24: The first wave began at 7:41 PM and lasted through midnight; the second wave launched during the pre-dawn hours of March 25, lasting from 1:00-3:30 AM.

During that time, the QRA MiG-29s were ordered to scramble. However, Serbian records confirm that only a handful of MiG-29s were available to meet the threat. Sadly, none of these MiG-29s would survive their encounters with NATO aircraft.

The first air-to-air engagement for a Serbian MiG-29 began shortly after 8:00 PM. This ill-fated Fulcrum, piloted by Major Iljo Arizanov, took off from Niš Air Base, vectoring towards Kosovo, patrolling airspace between Djakovica and Suva Reka. At the same time, Lieutenant Colonel Cesar "Rico" Rodriguez (the same pilot who had downed two Iraqi MiGs while assigned to the 58th Fighter Squadron in Desert Storm) took flight in his F-15 from Aviano Air Base. Now as a member of the 493d Fighter Squadron, he was flying as part of a four-ship formation for the inaugural foray into Yugoslavian airspace.

"There were two missions on this first night," Rodriguez recalled, "both spearheaded by F-15s of the 493d Fighter Squadron." Their intended targets were the radar and air defense missile sites in Montenegro, which Yugoslav forces had positioned to deny access to Pristina.

Heading north towards Montenegro, Rodriguez and his wingman, Lieutenant William "Wild Bill" Denim, found an initial radar contact 25 miles from a templated airfield. At first, it appeared to be an enemy air patrol, but its orbiting flight pattern and slow airspeed suggested that it wasn't a fighter patrol. While the flight leader focused on this enemy flight group, Rodriguez and Denim maintained their focus on Pristina. "Our main task was to ensure that nothing took off from Pristina heading towards Montenegro that might intercept our strike."

Suddenly, from a distance of 70 miles, Rodriguez got radar contact on a fast mover climbing to 10,000 feet. Using his onboard instrumentation, Rodriguez confirmed that this lurking bandit was a MiG-29 – Major Arizanov's plane. "I directed my element to start a climb, jettison tanks, and push it up. This would give our [missiles] greater range and since we were on the front edge of the strike package, I wanted to shoot as soon as possible…start the shooting match on our terms, not the MiG's."

Closing within 20 miles, Rodriguez fired an AIM-120 missile. Rodriguez hadn't realized it, but he had accelerated to Mach 1.3. Thus, as the missile glided off his wing, it appeared as though the AIM-120 was flying alongside him.

"The missile took a couple of seconds to build up momentum and accelerate out in front of me, and during that time, I thought I might have had a bad missile."

As the missile plowed forward into the night, Rodriguez could see the faint glow of the MiG-29's engines – "about the size of a dime," he recalled.

As he looked through his sights, Rodriguez lost visual contact of Arizanov's plane, as it was dark outside. But Rodriguez began counting down the seconds until the anticipated impact. When the counter reached zero, a huge fireball erupted. "Because the western mountains were still covered in snow," he said, "the fireball literally lit up the sky as it reflected off the snow-covered mountains. The only thing I had ever seen like this was when they turn on all the lights at an NFL stadium, except this was like five times as bright; it really lit up the whole sky."

Meanwhile, the Yugoslav Air Force realized that they'd lost the tracking signal from Arizanov's plane. They knew that his MiG-29 had either crashed or been shot down, but they had no immediate way of determining whether he had survived.

Two days later, however, Yugoslav ground forces found him, alive, trekking through the wilderness in Kosovo while evading the Kosovar insurgents. Although the MiG-29 had exploded in a violent fury,

Arizanov had successfully bailed out, parachuting into the frozen countryside, and had been making his way back to friendly territory.

When asked about the events leading up to his shootdown, Arizanov explained that both his radio and SPO-15 radar warning receiver had failed. Both of these instrumentation failures were attributed to the UN embargo which had halted the delivery of spare parts into Yugoslavia.

Naturally, Rodriguez's AIM-120 missile had taken Arizanov completely by surprise.

Moments before the shootdown, Arizanov recalled that he'd being trying to get "radar lock" onto another group of targets – presumably part of the incoming Allied strike package. He had barely identified the signature on his radar when, at 8:20 PM, Rodriguez's missile struck his MiG from the rear.

While Rodriguez closed in on Arizanov's MiG-29, Major Dragan Ilić took flight from Niš Air Base in his own MiG-29 at 8:12 PM.

Ilić was directed to patrol the airspace in the Metohija region near the Albanian border. During that patrol, his radar populated with the signatures of several enemy planes, though each of them was several miles away. At one point, while climbing to 14,700 feet, his MiG-29 was purportedly hit by an AIM-120 missile. Luckily for Ilić, he was able to keep his damaged Fulcrum in the air long enough to land safely at Niš around 8:45 PM.

To this day, however, it remains unclear if Major Ilić was truly hit by an AIM-120, or if he was the victim of "friendly fire" by a Yugoslavian SAM battery. For those who assert that Ilić was hit by the AIM-120, there is no clear consensus regarding the identity of the aggressing aircraft.

Meanwhile, as news of the incoming Allied aircraft began to spread, ground crews at Batajnica Air Base began setting fire to various wood pilings and tires, hoping to create a smokescreen against the anticipated airstrikes. Almost simultaneously, the Yugoslav Air Force sent two additional MiG-29s in the air: one piloted by Major Nebojsa Nikolic, the other piloted by Major Ljubiša Kulačin.

Nikolic took off at 8:37 PM, heading north towards Bečej. Soon into his flight, however, he discovered that his radar and missile tracking system had malfunctioned – another casualty of the under-resourced maintenance system. But little did he know that his flight path would put him in firing range of an F-15C piloted by Captain Michael Shower of the 493d Fighter Squadron.

Shower, like Rodriguez, had gone aloft with an assortment of other NATO aircraft – all of which were operating within a tight airspace. According to Shower, his mission was to fly as part of a two-plane formation and set up a Combat Air Patrol north of Belgrade. This would keep him out of the SAM envelope, yet still give him good coverage over the identified MiG bases.

Settling into his patrol, Shower heard the radio erupt with "Splash one MiG in the south!" – confirming Rodriguez's kill of the MiG-29.

Shower was excited.
He had been over Serbian airspace for all of seven minutes, and there was now enemy contact. His excitement intensified the moment he saw a radar contact of his own – 35 miles from his position, a fast-moving aircraft climbing up to 10,000 feet.

It was Nebojsa Nikolic's MiG-29.
Closing within 17 miles, Shower confirmed that the contact was an enemy bandit. Arming his missiles, he bellowed: "Hostile! Hostile! Fox 3!" – firing an AIM-120 from a distance of fourteen miles.

As soon as the AIM-120 left the missile rack, Shower thumbed his selector switch to AIM-7 and fired a Sparrow missile. However, both missiles failed to lock on to the fleeting MiG.

Now closing withing five and a half miles, Shower fired another AIM-120 and watched in amazement as the missile impacted the MiG-29.

Recalling the engagement years later, Shower didn't recall seeing the pilot eject, but he remembered watching the aircraft break apart and explode on the ground. However, he was happy to learn that Major Nikolic survived the encounter. In fact, Shower remarked that Nikolic, like himself, also had a wife and family waiting for his return.

After Nikolic ejected from his ailing MiG, he sent a

distress signal from the crash site, and in less than one hour, a Yugoslav heliborne search-and-rescue team retrieved him unharmed.

Meanwhile, aboard the MiG-29 piloted by Major Ljubiša Kulačin, instrumentation failures abounded. He had taken off only three minutes after Nikolic but, after climbing to nearly 10,000 feet, Kulačin discovered that his radar, too, was malfunctioning.

Frustrated by the ill-timed instrument failure, he spent the next several minutes trying to bring his radar back online. Kulačin managed to patrol *some* of the airspace near Bečej but, realizing that the broken radar had rendered him combat ineffective, he reluctantly decided to return to Batajnica Air Base.

By this time, however, the runway at Batajnica had been taken over by the smoke and fires set by the ground crews. With no visibility, Kulačin flew instead to Belgrade International Airport, where he landed without incident. Ironically, his MiG-29 remained parked and partially hidden among the flightline of commercial jets until the end of the war.

The fifth Serbian MiG-29 to go airborne that night was flown by Major Predrag Milutinović, part of the detachment at Ponikve Air Base. Milutinović took off at 8:45 PM to intercept the second wave of incoming airstrikes. Like most of his comrades, however, Milutinović was soon beset by malfunctioning comms and radar.

Luckily, his RWR was still functional, although it offered little solace as it gave him two proximity warnings. Deciding to let the threats bypass him, Milutinović landed at the nearby Ladjevci Air Base.

Because Ladjevci was under blackout conditions, Milutinović bought himself enough time to re-establish radio communication with his leadership. They ordered Milutinović to proceed to Niš Air Base, some 90 miles east. Milutinović acknowledged the order and set a course for what should have been a 10 to 15-minute flight. However, as he ascended to nearly 6,000 feet above Mount Jastrebac, Milutinović was hit by what he thought was anti-aircraft fire.

In reality, however, the culprit was a Dutch F-16. And it was the first time since World War II that a Royal Netherlands Air Force pilot had scored an air-to-air victory. That F-16, piloted by Captain Peter Tankik, was part of a four-ship detachment from the Dutch 322d Fighter Squadron at Amendola Air Base. That night, the Dutch F-16s had been flying cover for the US-led strike package. As they breeched Yugoslavian air space, they received an AWACS alert of three MiG-29s in the area – including Milutonović's plane. Moments later, Tankik had the signature of Milutonović's MiG on his radar. Locking on to the Fulcrum, Tankik fired his AIM-120. A mere thirty seconds later, the lurking MiG-29 disappeared from his radar screen. Milutonović, obviously startled by the sudden hit, ejected from the stricken plane. After safely parachuting to the ground, he was recovered by some local townsfolk and was taken to a nearby hospital in the city of Kruševac.

Meanwhile, the Yugoslav Air Force Chief of Staff, General Nikola Grujin, ordered the 83rd Fighter Regiment to release eight MiGs from the underground complex at Priština Air Base, and to put them on quick alert status along the flightline. For better or worse, the Yugoslav high command feared that the ongoing air strikes would hit the entrances to the underground hangars, thus preventing the reserve fighters from going airborne. Their premonition proved correct; merely two hours later, a flurry of Tomahawk cruise missiles descended on Priština. Impacting on the sloped sides of the underground complex, the blast was enough to precipitate an avalanche of earth and stone...all of which barricaded the southern entrance to the hangar. For the next several days, Yugoslavian ground crews enlisted a team of bulldozers to clear the wreckage.

During that first night of Operation Allied Force, the Yugoslav government counted three MiG-29s lost, and one severely damaged. General Wesley Clark, the Supreme Allied Commander in Europe, was elated by the news, but admitted to being shocked by NATO General Secretary Solana's reaction. Upon hearing news of the initial airstrikes and the downed MiG-29s, Solana reportedly shouted:

"That is no good! That is no good!"

Solana explained that these losses would draw NATO into a protracted conflict. Unlike the Iraqis, the Serbs were tenacious, adaptive, and weren't afraid to fight outnumbered. These air-to-air engagements proved that the Serb-Montenegrin forces not only wanted to keep Kosovo, they were willing to fight for it. According to Solana, the downed MiGs would surely provoke the Serbs into a war of attrition over Kosovo.

NATO politics aside, the air war continued unabated. On the morning of March 26, 1999, the Yugoslav Air Force lost another MiG-29 – this time to an accidental crash. At 9:30 AM, Major Slobodan Tešanović flew from Podgorica Air Base under orders to join the contingent at Ponikve. His MiG-29 had taken damage on the first night of the air campaign but, after some hasty repairs, it was deemed "airworthy" and ready for action. He took off from Podgorica and flew over western Serbia with no interference from Allied aircraft.

On his final approach into Ponikve, however, Tešanović succumbed to fatigue.

Having spent the last three nights on ready-alert status, he lost his bearings just a few hundred feet from the runway. Suddenly realizing that he'd lost control of the descent, Tešanović ejected while his Fulcrum crashed nearby and skidded to a halt beyond the runway. Tešanović survived the ejection, and there were no collateral deaths, but the Yugoslav Air Force was none too happy about the accidental (and likely preventable) loss of their aircraft.

Later that afternoon, two MiG-29s from the 204th Fighter Regiment at Batajnica went airborne, ready to intercept the next wave of Allied aircraft. The two pilots, Major Slobodan Perić and wingman Captain 1st Class Zoran Radosavljevic, initially took their air patrol over Bečej at a cruising altitude of nearly 10,000 feet. Major Perić, like many of his comrades, reported that his radar was malfunctioning.

Radosavljevic, however, had been unexpectedly blessed with a fully-functional radar.

Thus, the pair agreed that Radosavljevic would take command of the flight, and that he'd be the first to fire if they made enemy contact. Suddenly, they received a radio call alerting them to a pair of NATO aircraft west of their position.

As it turned out, the two radar contacts were American F-15s from the 493rd Fighter Squadron – piloted by Captain Jeff "Claw" Hwang and his wingman Lieutenant J. "Boomer" McMurray. At the time, Hwang and McMurray were eastbound, approaching the Bosnia-Serbia border, when they saw a radar contact 37 miles out, closing at 600 knots – Peric and Radosavljevic in their MiG-29s. Hwang recalled that the AWACS, surprisingly, couldn't see the same contact on their own radar.

Hwang and McMurray then vectored southward while continuing to monitor the new radar contact. He figured that this contact would probably continue southbound, or turn east while remaining south of the border. Still, Hwang alerted the AWACS of the contact and he passed the same report to his flight leader, who was conducting another aerial patrol over Sarajevo.

A few moments later, Hwang once again got radar contact on the incoming bogeys, who were now headed straight for them. From their relative position, Hwang could tell they were Yugoslav fighters. However, given the Rules of Engagement for today's mission, he needed clearance from the AWACS before firing. Hwang radioed the AWACS with the designated "code word" – indicating a request to engage a hostile target.

The AWACS, however, remained silent. At this point, Hwang was certain that the AWACS crew hadn't been briefed on today's "code word." However, once the target closed within 30 nautical miles, Hwang decided he wasn't going to wait for the AWACS's permission. Vectoring northeast, Hwang and McMurray gained visual contact on the bandits almost as soon as the AWACS confirmed that they were MiG-29s.

Closing within 16 nautical miles, Hwang fired his first missile.

He later recalled that his AIM-120 was so loud, he could hear the missile's roar even above the noise of the jet engines and radio chatter.

Hwang then fired his second missile.

"Fox 6! Lead Trail!" he cried as the missile glided

off its rail.

Diving down from 30,000 feet, Hwang acquired the trailing MiG in his sights. He was about to line up an AIM-9 missile shot when, suddenly, he saw the first plane explode – a victim of one of Hwang's first two missiles.

Seconds later, Hwang saw the trailing jet explode. Satisfied that both MiG's had been destroyed, Hwang un-selected the AIM-9, which he had intended to be his third missile shot of the afternoon. It was later confirmed that although McMurray fired first, it was both of Hwang's missiles that destroyed the respective bandits.

Both Peric and Radosavljevic ejected from their aircraft, but only Peric survived the ejection. Parachuting to the ground, Perić was apprehended by a group of Bosnian Serb villagers. But despite their common ethnicity, the villagers at first refused to believe that Peric was a Serbian pilot. Surprisingly, they thought he was a NATO pilot who could speak fluent Serbian. Peric finally convinced his "captors" that he was a true Serb by having them contact his cousin in Belgrade, who subsequently confirmed his identity.

As for Zoran Radosavljevic, his plane crashed south of Bijeljina. According to the Serbian government, Radosavljević's body was still in the ejection seat when it was discovered by two teenage boys later that evening. The fuselage was found in a nearby field, while the nose section landed on the slope of a nearby mountain. The teens left the wreckage undisturbed, but wrapped Radosavljević's body in a blanket before handing him over to the Army of Republika Srpska. The Serb operatives then expedited him to the cross-border hospital in Loznica. Radosavljevic was buried three days later at Lešće Cemetery in Belgrade.

Although the first month of Operation Allied Force was a resounding success for NATO's air forces, they did not emerge completely unscathed. For instance, on March 27, 1999, an F-117 Nighthawk (the vaunted "Stealth Fighter") from the US 8th Fighter Squadron was shot down 25 miles northwest of Belgrade. Although the aircraft bore the name "Capt. Ken Dwelle" on the fuselage, the F-117 was actually being piloted by Lieutenant Colonel Dale Zelko on the night of the mission. Luckily, Zelko ejected from the downed Nighthawk and was safely recovered by an Air Force Pararescue team.

The culprit was later identified as an SA-3 air defense missile fired by the Yugoslav Army's 3rd Battalion, 250th Air Defense Missile Brigade, commanded by Colonel Zoltán Dani. However, in May 1999, Russian media began circulating rumors that the F-117 had been downed by a MiG-29. *Krasnaya Zvezda*, the Russian Ministry of Defense news service, reported that a Yugoslav MiG-29 flown by Lieutenant Colonel Gvozden Djukic had fired on the F-117 while flying a combat air patrol along

Serbia's western border. Russian media further stated that Djukic had been awarded a flying medal from Yugoslav President Slobodan Milosevic. The fantasy article even quoted Djukic saying: "I felt as pleased as an Indian who had just taken a scalp, or even more." In reality, the Yugoslav Air Force had no such pilot named Gvozden Djukic. To the contrary, "Gvozden Djukic" was simply Colonel Zoltan Dani's *nom de guerre*. As mentioned earlier, Zoltan Dani was the commander of the SAM battalion that had downed Zelko's F-117.

On May 4, 1999, the Yugoslav Air Force suffered its final Mig-29 loss. It was arguably the most devastating loss of the war. Indeed, the ill-fated Fulcrum was piloted by Lieutenant Colonel Milenko Pavlovic, commanding officer of the 204th Fighter Regiment. As stated by prominent Serb historians Bojan Dimitrijevic and Jovica Draganić, the death of Lieutenant Colonel Pavlovic was a "decisive moment, after which Yugoslav Air Force fighters ceased to operate until the end of the campaign." He was regarded by many as the best pilot in Yugoslavia.

At just past noon on May 4, Lieutenant Colonel Pavlovic was at Batajnica when he heard the news that his hometown of Valjevo had been bombed by NATO aircraft. According to the latest intelligence, these same aircraft were now headed west attempting to exfiltrate Yugoslav airspace. Consumed by vengeance and rage, Pavlovic commandeered one of the QRA MiG-29s, taking off at 12:37 PM. He hastily set a course to Valjevo, some 60 miles southwest of Belgrade. Climbing to 10,000 feet, Pavlovic avoided the occasional bursts of "friendly fire" from Yugoslav air defense batteries on the ground.

Establishing contact with the nearest radar surveillance unit, Pavlovic got his bearing onto a flight of egressing US aircraft. While accelerating to engage the NATO bandits, however, Pavlovic's MiG populated on the radar of an American F-16 piloted by Lieutenant Colonel Michael "Dog" Geczy from the 78th Fighter Squadron. Locking onto Pavlovic from beyond visual range, Geczy brought down Pavlovic with a single AIM-120.

At 12:47 PM, Yugoslav air surveillance lost the signal from Pavlovic's MiG.

Barely one-half hour later, the Yugoslav Air Force confirmed that Lieutenant Colonel Pavlovic, Commander of the 204th Fighter Regiment, had been killed in action. The news sent shock waves throughout the Yugoslav military. To that point in the war, Pavlovic was the highest-ranking Yugoslav commander to be killed in action; and his untimely death represented the *sixth* MiG-29 destroyed in the air war over Kosovo. The Yugoslav Air Force consequently suspended all fighter activity for the remainder of the conflict.

Operation Allied Force ended on June 10, 1999. Per

the terms of the Kumanovo Agreement and UN Resolution 1244, Yugoslav forces withdrew from Kosovo and the region was granted full autonomy, if not partially-recognized sovereignty. The Yugoslav Air Force, however, had taken a tremendous beating. By the end of the war, they had lost nearly 50 aircraft, 40 airmen killed in action, and 110 wounded.

After the war, only five MiG-29s remained in service. However, their flight hours were drastically reduced due to the ongoing supply shortage. In fact, the Yugoslav Air Force itself would cease to exist in 2003. When the post-Milosevic government dropped its claims of successorship to the Tito era, the Republic of Yugoslavia voted itself into the latter-day confederation of "Serbia and Montenegro." In keeping with its new identity, the Yugoslav Air Force then became the Air Force of Serbia and Montenegro.

The name change, however, did nothing to alleviate the ongoing readiness issues. By 2004, reports had surfaced that Serbia and Montenegro had grounded its entire fleet of MiG-29s, as they no longer had the means to keep them flying. Two years later, Montenegro seceded from its union with Serbia, thus leaving the two entities fully independent, and formally dissolving the last remnants of the Yugoslavian identity.

Following their dissolution, the new Serbian Air Force retained the bulk of its predecessor's inventory, including the grounded MiG-29s. Rather than scrap the five remaining Fulcrums, however, Serbia sent them to Russia for an extensive overhaul. Notwithstanding the 2009 loss of a MiG-29 in a crash over Belgrade, these handful of re-fitted Fulcrums remain in service. Starting in 2017, however, Russia and Belarus donated fourteen newer MiG-29s to the Serbian Air Force. Shortly thereafter, Serbia announced its plan to spend nearly €200 million to modernize its fleet of MiG-29s. As of 2022, all fourteen of these MiG-29s are still flying in the Serbian Air Force.

A post-war Serbian MiG-29 featuring the R-60 (AA-8 Aphid) missiles, 2011. After the Kosovo War, and subsequent dissolution of the Serbian-Montenegrin political union, Serbia's Air Force retained its legacy fleet of FR Yugoslav MiG-29s. More recently, Serbia has announced a modernization program to keep its Fulcrums on par with the current generation of military avionics. *Krasimir Grozev*

A Syrian MiG-29 prepares for takeoff during the Syrian Civil War. The latter-day Syrian Air Force was the aerial warfare branch of the al-Assad regime. Syrian MiGs often flew combat patrols in loyalty to the al-Assad government. However, when Damascus fell to the Syrian Opposition in December 2024, the entire Syrian Air Force (including its fleet of MiG-29s) was destroyed during a simultaneous air raid led by the Israeli Defense Forces. *Zvezda/Russian Ministry of Defense*

A Yemeni MiG-29, ironically nicknamed "Shafaq." In Arabic, the word "Shafaq" translates literally to "twilight," but colloquially, it can also mean "affection" or "sympathy." *Yemeni Majlis al-Nuwaab*

Chapter 5:
Brushfire Wars & The Arab Spring

India

From its inception in 1947, the Republic of India had maintained good relations with the United States and the Soviet Union. After gaining its independence from Britain, India had no stake (and little interest) in the broader politics of the Cold War. As it turned out, the Indian government had more pressing matters at hand – mostly involving Pakistan and Maoist China. To that end, India graciously accepted military sales from both sides of the Iron Curtain.

Beginning in 1987, the Indian Air Force (IAF) took delivery of forty MiG-29 9.12B single-seat variants and four MiG-29UB two-seat trainers. Satisfied by their initial performance, the IAF ordered an additional twenty-six aircraft in 1989, followed by ten MiG-29 "9.13" upgraded export variants in 1994. Since that time, India's land-based MiG-29s have undergone a series of domestic and foreign-sourced modifications – including upgrades to the avionics suite, radar system, and powerplant.

Nicknamed "Baaz," (Hindi for "eagle" or "hawk"), the IAF MiG-29s have routinely patrolled the frontiers of their domestic airspace along the Himalayan Mountains and in the Vale of Kashmir. But in May

1999, while Serbian MiG-29s were on the defensive in Operation Allied Force, the IAF Fulcrums made their combat debut during the brief-but-intense Kargil War.

India and Pakistan had been bitter rivals since their partition during the final days of the British Raj. In fact, throughout the second half of the 20th Century, India and Pakistan repeatedly clashed over territorial issues. These ranged from small-scale firefights along the border to full-scale military conflicts like the Indo-Pakistani Wars of 1947, 1965, and 1971. One of the flashpoints in their ongoing territorial struggle was the latter-day principality of Kashmir. Following the latest war in 1971, the Simla Agreement divided Kashmir along the so-called "Line of Control" (LoC), creating a *de facto* border between Indian and Pakistani territory.

However, in the summer of 1999, Pakistani troops disguised as Kashmiri guerrillas infiltrated the Indian side of the LoC, initiating a series of attacks within the Kargil district. While these Pakistani operatives had hoped to maintain their disguise as Kashmiri guerrillas, the Indian government soon discovered proof of Pakistan's involvement. This, in turn, prompted the Indian government to eject the Pakistanis from Kashmir by force.

By the time hostilities began on May 3, 1999, the IAF had been flying regular reconnaissance patrols over the area. For the upcoming strikes against Pakistani ground forces, however, the IAF alerted their attack squadrons from three regional bases: Srinagar, Avantipur, and Udhampur. Taken together, these squadrons provided rotational strike teams of MiG-21s, MiG-23s, and MiG-27s...with MiG-29s providing top cover.

As it turned out, the IAF MiG-21s and MiG-27s would carry most of the burden for the air campaigns in Kargil. During the opening rounds, IAF MiGs and helicopter gunships provided close air support for the Indian Army as part of Operation Safed Sagar – a coordinated air-ground assault against Pakistani strongholds. IAF Captain Alok Chauhan, a retired MiG-21 pilot, summarized the early Kargil strike missions as follows:

> "Our job was to go along with the strike aircraft like the MiG-23s and MiG-27s who were armed with bombs and rockets and provide them air defence cover in their immediate vicinity. The MiG-29 provided us an additional bubble of cover since they had more potent fire power and stand-off ranges."

Simultaneous to the ground attack missions, IAF Western Air Command sent their MiG-29s to occupy combat air patrol (CAP) stations at the forward edge of the LoC. These CAP stations allowed the Indian Fulcrums to keep tabs on Pakistani troop movements and intercept any prowling fighters from the Pakistan Air Force (PAF). The PAF, much like the IAF, had also maintained a stable relationship between East and West. But Pakistan had received the coveted F-16 Fighting Falcon – one of the best multi-role fighter jets in the world.

Despite the Pakistanis' aeronautical advantage, however, the IAF had strength in numbers. IAF fighters outnumbered the PAF nearly 2-to-1. When comparing their top-rated fighters, the PAF had only twenty-six F-16s against seventy Indian MiG-29s.

To the disappointment of aerospace analysts worldwide, the much-anticipated F-16 vs MiG-29 dogfight never came to be. Instead, Pakistani F-16s typically stayed within 10-20 miles of the LoC.

There were, however, a few instances of IAF MiGs locking on to Pakistani F-16s from across the border. On one such occasion, Lieutenant Garauv Chibber, the youngest MiG-29 pilot in the IAF, saved his fellow airmen from an F-16 intercept. On that day, an IAF MiG-27 strike package reportedly strayed across the LoC, whereupon they were quickly detected by a two-ship flight of Pakistani F-16s. The marauding F-16s vectored to intercept the MiG-27s; but Chibber, flying cover in his MiG-29, swooped in and locked on to the incoming bandits with his radar. The F-16s, now realizing that they were on Chibber's radar, hastily broke off the intercept.

He was commended for his actions that day; but shortly after the war ended, Chibber died in August 1999 when his MiG-29 crashed on a routine flight over Himachal Pradesh. Chibber's death (alongside his comrade, Flight Officer Pankaj Joshi, who perished in a similar crash the following December) prompted the IAF to review its safety/maintenance program for the MiG-29. In fact, the deaths of Chibber and Joshi brought the total number of that year's non-combat MiG-29 fatalities in the IAF to 28. It was an alarming number for any air force of its size.

By the end of the summer, the Indian Army had recaptured much of the occupied territory in Kashmir. The Pakistani government, meanwhile, bowing to international pressure, withdrew its forces from the Indian side of the LoC. Although the Indian government had declared victory, the IAF had not emerged from the conflict unscathed. Their losses included some high-profile shootdowns of a MiG-21 and a MiG-27. These losses prompted the IAF to modernize their fighter fleet, and precipitated development of the Su-30MKI, India's licensed copy of the Sukhoi Su-30 multirole fighter.

Safety issues notwithstanding, the MiG-29 Baaz received high praise for its performance in the Kargil War. In light of its operational record, the Indian government brokered a deal with Mikoyan in 2005 to upgrade the IAF's entire fleet of MiG-29s. At a cost of nearly $900 million (USD), Indian MiGs received the new Phazotron radar, an improved avionics suite, and modifications to accommodate the new R-77 air-to-air missile. Further enhancements included a new weapons control system, improved ergonomics within the cockpit, and improved armaments for air-to-ground operations.

At around the same time, India ordered its first contingent of MiG-29K carrier-based fighters for its growing arm of naval aviation. Initial deliveries began in December 2009, with the naval-variant Fulcrums entering operational service in February 2010. Since their debut with the Indian Navy, however, the MiG-29Ks have been beset by numerous safety concerns. For example, a 2016 naval audit revealed that the operational readiness rate for India's MiG-29K fleet hovered between 15-37%, citing numerous defects within the engine suite, airframe, and onboard avionics. Moreover, recent international sanctions against Russia have disrupted the flow of supplies to keep India's MiG-29Ks in top condition.

Although the land-based and naval-variant MiG-29s remain in service with the Indian military, there have been reports since 2017 that the IAF plans to retire the MiG-29, eventually replacing it with a Western-based multirole fighter.

Eritrean-Ethiopian War

In the realm of aviation history, 1999 was truly the "Year of the Fulcrum." Overlapping the Kargil War and Operation Allied Force, the MiG-29 saw extensive use during the border conflict between Ethiopia and Eritrea.

Since gaining its independence from Ethiopia in 1991, Eritrea had been embroiled in an ongoing territorial dispute with its former parent country.

Before the schism, however, Ethiopia had been among the many Soviet client states in Africa. Throughout the 1970s-80s, the USSR had supplied Ethiopia's Air Force with a full contingent of MiG-17s, MiG-21s, and MiG-23s.

The new Eritrean Air Force, however, hadn't been so lucky.

Only a handful of MiG-21s captured from Ethiopia formed the backbone of Eritrea's fighter corps. When hostilities finally erupted between the two African states, Russia happily sold weapons to both sides. Ethiopia bought the newer Su-27 "Flanker," to which the Eritreans responded by purchasing MiG-29s. Meanwhile, a cadre of Russian (and Russian-speaking) advisors, instructors, and mercenary pilots arrived in both countries. For example, the Eritrean Air Force procured and organized its MiG-29 fleet with the help of Colonel Vladimir Nefyodov, a freelance arms dealer who had arranged the delivery of Fulcrums to Eritrea direct from Mikoyan, thus bypassing Russia's state-controlled arms distributors.

An initial delivery of six MiG-29s arrived in Eritrea in November 1998. In tow was a cadre of contracted Ukrainian instructor pilots and technicians, along with several dozen R-27 and R-73 missiles. The Eritrean Air Force had only *four* pilots available to operate the incoming Fulcrums, all of whom received a two-week conversion course in Russia. One of the Eritrean pilots, Dejen Ande Hishel, recalled that:

"Our training was undertaken in great hurry and

entirely inadequate. I was shocked when given a certificate from the Russians stating not only that I was a qualified pilot, but also a qualified instructor pilot. Having only been taught the fundamentals, we never logged the sufficient number of flight hours, never trained formation flying or for combat, and thus we had no confidence in our abilities. I protested my premature graduation, but to no avail. My conclusion was that the Russians were more interested in meeting the goals of the higher-ups [i.e. a quick turnaround for graduates] than meeting our goals."

After returning to Eritrea, Dejen and his colleagues discovered that their MiG-29s were older models, and that they frequently malfunctioned. Considering the cyclic rate at which their avionics failed, the pilots began to wonder if their ejections seats would even work.

But given the circumstances, there was little they could do.

The ill-equipped Eritrean pilots thus prepared to launch their ailing MiG-29s, hoping they could maintain enough standoff distance to fire their R-27 missiles against the Ethiopian fighters.

On February 5-6, 1999, the Eritrean MiG-29s flew their inaugural combat missions against the Ethiopian Air Force. Responding to Ethiopia's air and artillery strikes, two Eritrean MB-339s bombed a fuel depot in

Adigrat. Meanwhile, the Eritrean MiG-29s fired their R-27 missiles at a flight of Ethiopian helicopters and MiG-23s, but none of the missiles hit their intended targets. During this engagement, the Ethiopian Air Force scrambled at least one Su-27 to intercept, but the pilot arrived too late to catch the offending MiGs.

The following day, the Eritrean Air Force tried to lure the Su-27s into an ambush, but the attending MiG-29s were soon distracted by a pair of "bait" MiG-21s. One of the Fulcrums got close enough to the Ethiopian MiG-21s to fire an R-27 missile from maximum range.

Although this missile, too, failed to hit its target, it vectored close enough to the Ethiopian MiGs that the lead pilot could hear the missile detonate behind his aircraft. The Eritrean Air Force may have been undertrained and underestimated…but for the Ethiopians, this engagement was too close for comfort. A startled pair of nearby Su-27s returned fire with a volley of their own R-27s, hoping to hit the elusive Fulcrum that had nearly downed the baiting MiG-21, but their missiles vectored off, and the Fulcrum soon disappeared from radar view.

On February 21, a pair of Eritrean MiG-29s attempted to ambush another Su-27 by way of a pincer attack. However, because the Eritrean radar station at Adi Quala had been destroyed days earlier, one of the MiG-29s inadvertently flew into an Ethiopian ambush. Once again following a bait-trap MiG-21, the Eritrean MiG-29 climbed to nearly

20,000 feet, whereupon it was ambushed by a loitering Su-27, firing its first missile from a distance of nearly 30 miles. Luckily, the first missile veered off target, prompting the Eritrean MiG to turn away. The Ethiopian pilot, however, gave chase and fired his second R-27 missile from a range of about six miles. This time, the R-27 detonated its proximity fuse, sending the MiG-29 to its death 20,000 feet below.

Moments later, the other MiG-29 re-emerged to engage the Ethiopian bandit, firing his own R-27. However, because the enemy bandit was already accelerating to supersonic speed, this missile, too, failed to hit its mark.

Four days later, MiG-29 pilot Yonas Misghinna scrambled from Asmara Air Base to intercept a reported Su-27 north of Mekelle. Despite his very deliberate and low-altitude approach, Misghinna's MiG-29 populated on the radar of another Su-27, piloted by Ethiopian Lieutenant Colonel Gebre-Selassie. Vectoring onto Misghinna's position, Gebre-Selassie fired two R-27 missiles, one of which scored a direct hit, sending Misghinna's MiG-29 hurtling into the desert floor.

By this time, the Eritrean Air Force had only *two* MiG-29 pilots remaining. Dejen Ande Hishel, one of the two surviving Eritrean MiG pilots recalled:

> "I was sad and angry and filled with the wish to avenge Yonas' death. Normal procedure would

have been to run an investigation and find out why he was shot down, in order to avoid similar mistakes in the future. In the meantime, we should have avoided repeating the pattern of the circumstances that led to the loss, which means flying over the same area, altitude and speed of the downing. However, our government ignored all of this and ordered us into the sky again."

The following day, February 26, Flight Officer Samuel Girmay scrambled his MiG-29 from Asmara, vectoring south to meet an unspecified airborne threat. The outcome of Girmay's engagement is undisputed: his MiG-29 was downed by Lieutenant Colonel Gebre-Selassie, the same Su-27 pilot who had downed Misghinna's Fulcrum the day before. But the details of Girmay's engagement remain unclear.

Indeed, there are *three* different versions of the story.

According to one version, Gebre-Selassie shot down the MiG-29 from beyond visual range, firing a single R-27 missile. In another version of the story, Girmay and Gebre-Selassie exchanged fire from standoff distances; the Ethiopian Air Force claimed to have intercepted a radio transmission from Girmay announcing his missile shot. In the third version of the story, Gimray and Gebre-Selassie exchanged multiple volleys from shorter distances. According to this version, Gebre-Selassie fired two R-27s, both of which failed to hit the MiG-29. Gimray then closed with the Su-27 and returned fire with his own R-27 missile. During this exchange of fire, however, Gebre-Selassie fired his heatseeking missile, which made first impact, killing Gimray and allowing the Ethiopian Flanker to evade the oncoming R-27.

With these back-to-back engagements, Ethiopian Su-27s had dealt a devastating blow to the Eritrean Air Force. With only two MiG-29s remaining (and two rated pilots left to fly them), it appeared that Eritrea's days were numbered.

Still, the Eritrean Air Force refused to quit.

Throughout 1999, Brigadier General Habtezion Hadgu, Commander-in-Chief of the Eritrean Air Force, worked hard to keep his dwindling fleet of MiG-29s operational. With his cadre of ex-Ethiopian personnel and contracted Ukrainian instructors, Hadgu spent considerable time training a second group of Eritrean pilots to fly the MiG-29. "His efforts were hampered by the fact that only two aircraft were left in operational condition, and that the stock of ammunition and spares for them were meanwhile critically low. The government in Asmara did attempt to obtain additional aircraft, spares and armament, and corresponding requests were forwarded to Moscow."

The Russian government responded favorably, but did not deliver the new MiG-29s until after the war ended.

By May 2000, as the Ethiopian military launched

its final offensive, the Eritrean Defense Forces were struggling to survive. By then, the "air war" had devolved into a mishmash of ground attack missions and near-misses with air-to-air missiles.

In mid-May, the Ethiopian Air Force lost an Mi-24 helicopter that crashed near Gefersa, killing its crew and their Russian advisor. Another helicopter sustained damage from Eritrean ground fire, but its crew managed to land the ailing gunship behind friendly lines. Rumors abounded that an Ethiopian Su-27 shot down another MiG-29, but neither side has ever confirmed the engagement. Moreover, the name of the alleged victim – a MiG pilot identified only as "Major Workneh" – does not appear on any of the Eritrean defense rosters.

According to aviation historians Tom Cooper and Adrien Fontanellaz, these reports were likely caused by the "sighting of a lone Eritrean MiG-29 in the skies west of Asmara during an air raid by two MiG-23BNs against the port of Massawa." By this point in the war, Ethiopian Su-27s were regularly escorting MiG-23s into battle. However, the Flanker pilots eventually stopped engaging Fulcrums because the Ethiopian Air Force's priority was to ensure the safe return of the MiG-23s.

The following month, the Eritrean-Ethiopian War ended via the UN-mediated Algiers Agreement. It was a military victory for Ethiopia, but a diplomatic victory for Eritrea. The UN's Boundary Commission concluded that the disputed territory did, in fact, belong to Eritrea. The ensuing peace was tenuous, however, until June 2018, when Ethiopia's Prime Minister Abiy Ahmed, agreed to fully implement the UN-mediated peace treaty of 2000.

The Eritrean Air Force was left in disarray by the conflict, but has been steadily rebuilding since the early 2000s. Reviewing the performance of their MiG-29s, there was little debate that the R-27 missiles had performed poorly in combat. On both sides of the conflict, the R-27 had barely achieved a 16% kill ratio. Postwar analyses, however, attributed these low performance numbers to "rough handling during transport, harsh storage conditions, and ill-trained weapons personnel." Of course, the slapdash training program given to the Eritrean MiG pilots didn't help, either.

After the war, Eritrea ordered six additional MiG-29s from Moldova. Likewise, the Eritrean Air Force received the Su-27s previously promised from Russia, but reports indicate that only two of these Flankers are actively in service.

At this writing, Eritrea has 11 MiG-29s in active service, but only eight are known to be flyable. For nearly a decade, the Eritrean government employed Russian defense contractors to help maintain their fleet of MiGs and Sukhois. However, this agreement ended in 2009 after the UN Security Council passed Resolution 1907, creating an arms embargo against Eritrea for its alleged support of Somali terrorists and the occupation of Djibouti. Thus, it appears that many of the Eritrean MiG-29s will remain grounded until the embargo is lifted.

Sudan

As a footnote to the Eritrean-Ethiopian War, the Sudanese Air Force followed a similar path in its wars against the Darfur insurgency, the breakaway Republic of South Sudan, and the more recent Sudanese Civil War.

The Sudanese Air Force, considered among the best in Africa, received much of its present-day inventory from China and the former Soviet Union. At this writing, the Sudanese Air Force has an estimated ten MiG-29s in service. These Fulcrums, purchased from Russia throughout the 2000s, form the backbone of Sudan's present-day fighter fleet.

In May 2008, during the throes of the Darfur insurgency, one of the rebel groups known as the Justice and Equality Movement (JEM) stormed the Sudanese capital of Khartoum. During this assault, JEM operatives downed a Sudanese MiG-29 with heavy machinegun fire while it was attempting to strafe an insurgent vehicle convoy.

Ironically, the pilot wasn't Sudanese; he was Russian.

Like many of their African counterparts, Sudan has employed a number of Russian mercenaries to pilot their Soviet-made aircraft. The pilot was killed when his parachute failed to deploy after ejecting. Russian sources did not identify the fallen aviator by name, but confirmed that he was a retired Soviet Air Force pilot. The Russian media further acknowledged that mercenary pilots had been flying Sudanese MiG-29s since their arrival in Khartoum.

A few years later, Sudanese MiG-29s were back in action during the Heglig Crisis of 2012. The Republic of South Sudan had been sovereign for less than a year when a border dispute erupted with the former parent state of Sudan. Not surprisingly, the oil-rich regions of the Sudanese border (mostly near the town of Heglig) became the epicenter of this land dispute.

In April 2012, as Sudanese forces sparred on the ground, their Fulcrums took to the sky. Although South Sudan had only a nominal "air force," their land component – the Sudan People's Liberation Army (SPLA) – was fairly well-equipped. On April 4, the SPLA claimed to have shot down a Sudanese MiG-29 with anti-aircraft fire.

The Sudanese government, however, quickly denied the incident.

Military spokesman Al-Sawarmi Khalid went on the record to say: "There were attacks by some groups on our military positions and we responded with artillery. Claims that a plane was shot down are non-befitting since we haven't used planes they claim to have shot down."

But even if the Sudanese MiGs weren't airborne on

April 4, South Sudanese intelligence (and various news outlets) confirmed their presence during the airstrikes of April 14 and April 22 near the oil town of Bentiu.

During the raid of April 14, two MiG-29s attempted to destroy a bridge in Bentiu, but instead accidentally killed four civilians and one soldier, while wounding five additional soldiers. The botched attack was intended to disrupt the South Sudanese supply lines. Following the airstrike of April 22, South Sudan's military intelligence corps reported that Sudanese MiG-29s had dropped at least four bombs in the area surrounding Bentiu. A Reuters correspondent on the ground during the attack confirmed seeing a "fighter aircraft drop two bombs near a river bridge between Bentiu and the neighboring town of Rubkona," presumably one of the same MiG-29s mentioned by South Sudanese intelligence.

The border conflict lasted barely six months, with Sudan being the clear winner. Under mediation from the African Union, Sudan gained control of the contested area while South Sudan's forces withdrew from Heglig. The Sudanese Air Force, meanwhile, continued operating its fleet of MiG-29s, Mig-23s, and Su-24s.

Although no Sudanese MiG-29s participated in the Yemeni Civil War (of which Sudan was a member in the Saudi-led interventionist coalition), their MiG-29s have played a prominent role in the ongoing

Sudanese Civil War. Beginning in April 2023, two competing factions within the military government of Sudan – one led by General Abdel Fattah al-Burhan; the other led by General Mohamed "Hemedti" Dagalo – erupted into an armed conflict.

On April 15, 2023, the first day of the civil war, a Sudanese MiG-29 was filmed firing air-to-ground missiles aimed at Hemedti's paramilitary fighters, the Rapid Support Forces (RSF), operating in Khartoum. However, in a dramatic turn of events, the RSF successfully downed a Sudanese MiG-29 over Omdurman on May 25.

Prior to that shootdown, RSF guerrillas had surprisingly captured three *Egyptian* MiG-29s when the paramilitary group stormed Merowe Airbase, 185 miles north of Khartoum. Satellite imagery confirmed that two of these Egyptian MiG-29's had been damaged, while the third had been destroyed. Almost simultaneously, RSF ground forces captured a number of Egyptian troops in the vicinity of Merowe. Initially, Egypt's government fell under suspicion for aiding al-Burhan's Sudanese forces. However, the RSF accepted Egypt's explanation that their personnel had been on a training exercise in Sudan when the civil war erupted. A few days later, the captured Egyptian troops were released and flown back to Cairo.

Most recently, in April 2024, the Algerian Defense Ministry announced that they would be donating their surplus MiG-29s to the Sudanese Air Force. At present, the Algerian Air Force has nearly 40 MiG-29s in service, most of which came from Russia and Belarus. The Algerian Defense Ministry had planned to phase-in the upgraded MiG-29SMT fighters, but opted instead for the Sukhoi Su-30MKA multirole fighter. As the newer Su-30s begin forming the backbone of Algeria's fighter fleet, the remaining Fulcrums are set to be refitted and transferred to Sudan for use in the ongoing civil war.

Georgia

The Russian Air Force did not deploy its MiG-29s during the Chechen Wars (1994-2009). However, the brief Russo-Georgian conflict in 2008 marked the first time that post-Soviet MiG-29s flew in a wartime mission. A former Soviet republic, Georgia declared its independence from the USSR in April 1991.

However, like many of the former constituent republics, Georgia was suddenly beset by a slew of ethnic and nationalistic divisions from within. By 1992, the ethnic Abkhaz and Ossetian minorities in Georgia had declared their intent to secede. After a series of brief, bloody civil wars, the republics of Abkhazia and South Ossetia achieved partial recognition from the international community (and a begrudging acknowledgment of autonomy from Georgia).

However, when Georgian President Mikhail Saakashvili won reelection in January 2008, he declared his intention to reintegrate Abkhazia and South Ossetia into the Georgian mainland. Meanwhile, Georgia's relationship with Russia continued to deteriorate, as the latter supported South Ossentian independence.

Tensions continued to rise after April 20, 2008, when Georgian officials claimed that a MiG-29 shot down a reconnaissance drone near the Abkhaz border. Onboard video footage from the Georgian UAV showed an apparent MiG-29 firing its missile squarely at the camera. Russia was quick to deny the allegation, claiming that they had no MiG-29s in the air that day. UN investigators concluded that the video was authentic, but couldn't determine whether the offending aircraft was a MiG-29 or Su-27. Forensic analysis did conclude, however, that the weapon of choice was likely an R-73 missile. Nevertheless, Georgia held fast to its claim that since neither they nor the Abkhazians flew MiG-29s, the culprits must have been Russian.

Finally, on August 1, 2008, South Ossetian forces launched an artillery attack against the Georgian border towns. In response, Georgian ground units stormed into South Ossetia and, by August 7, they had taken control of Tskhinvali, a separatist stronghold. However, Russian Ground Forces had preemptively crossed the Georgian border through the Roki Tunnel, and were lying in wait for the Georgian Army by August 7.

The following day, Russia launched a full-scale

invasion of Georgia, justifying its actions as a safeguard against Georgia's "aggression" towards South Ossetia.

In the sky, Georgia could mount little resistance as they had no fighter aircraft to match the Russian Air Force. The Georgian air defense command did, however, scramble their Su-25 attack planes for a series of strike missions against enemy ground troops on the first day of the war. Meanwhile, Russian MiG-29s from the 19th and 31st Fighter Regiments went aloft escorting the Beriev A-50 AWACS to monitor Georgian airfields. Beyond this fighter escort role (and the alleged drone "shootdown"), Russian Fulcrums played no further role in the conflict.

Over the next several days, however, Russian and South Ossetian ground troops put Georgian forces on the retreat. Russian-Abkhaz forces then opened a second front by attacking a Georgian stronghold in the Kodori Gorge. At the same time, Russian naval forces blockaded the Georgian sector of the Black Sea coast. Hostilities ended on August 12 when Russian President Dmitry Medvedev and Nicolas Sarkozy, the President of France, negotiated a ceasefire. Finally, on August 16, all four belligerents – Russia, Georgia, Abkhazia and South Ossetia – signed the peace agreement. Russian troops began withdrawing from Georgia later that month, but kept a small military presence in South Abkhazia to deter any future action from Georgian ground forces.

Syria

Like most of their Arab neighbors, the Syrian Air Force had relied heavily on Soviet and Russian equipment. After years of maintaining their fleet of MiG-23s and MiG-25s, the Syrian Air Force received its first delivery of MiG-29s in 1987. At the time, Syrian forces still occupied much of Lebanon, and Syrian MiG-23s had frequently traded fire with Israeli F-15s near the Lebanese border.

As the new MiG-29s came into service, they, too, would encounter a number of Israeli fighter and reconnaissance aircraft. Although the IDF has confirmed *seeing* MiG-29s along the edges of its territorial airspace, it remains unclear whether any Israeli aircraft have truly *engaged* the Arab Fulcrums.

Purportedly, on June 2, 1989, two Israeli F-15Cs downed a pair of Syrian MiG-29s following a brief dogfight. However, the IDF has never confirmed nor denied the incident. The names of the alleged pilots have never been released; no wreckage has ever been found; and there are no details regarding the location of the alleged incident.

In a similar vein, Syria and Israel have both denied reports of a similar "shootdown" that allegedly occurred on September 14, 2001. In a story published by the *World Tribune*, an Israeli Air Force Boeing 707 reconnaissance plane was on a mission patrolling the Lebanese and Syrian coast, escorted by two F-15s. By now, the Syrian Air Force had grown accustomed to the IDF's aerial recon missions, and typically scrambled their MiG-23s or MiG-29s to shadow and monitor the Israeli spy planes from distances of 12-13 miles. This time, however, two MiG-29s appeared on an aggressive vector, increasing their speed towards the Israeli aircraft. Sensing the MiGs' hostile intent, the lead F-15 pilot ordered the 707 to vector away and engage its electronic countermeasures. Moments later, he purportedly warned the Syrian MiG-29s via the international distress frequency to change their course. When the MiGs failed to respond, however, the F-15s moved in to attack, each downing a respective bandit with a missile shot. The Syrian pilots were identified as Major Arshad Midhat Mubarak and Captain Ahmad al-Khatab, both of whom purportedly ejected and were later recovered by Syrian ships. However, the alleged IDF pilots were never identified, and both sides still deny that this 2001 incident ever happened.

However, there is no dispute that the MiG-29 has been an active participant in the Syrian Civil War. Beginning in March 2011, as part of the wider "Arab Spring" movement, popular protests against Syrian leader Bashar al-Assad led to large-scale government crackdowns. This, in turn, precipitated the rise of various armed rebel groups, including the Free Syrian Army. By the fall of 2012, the Syrian insurgency had escalated into a full-blown civil war.

Ironically, the Syrian Civil War soon devolved into

its own "war by proxy" involving the latter-day Cold War powers. NATO backed the rebel groups; while Russia (and Iran) supported al-Assad's regime.

Initially, however, the Syrian government relied on its own forces to quell the rebellion. To this end, the Syrian Air Force deployed a number of MiG-29s alongside its existing fleet of Soviet-built fighter-bombers, including the Sukhoi Su-22. As a general rule, the ground-attack Fulcrums strafed rebel positions using unguided non-precision rockets, leading to collateral damage and further erosion of internal support for the al-Assad regime.

Almost simultaneous to the Arab Spring protests in 2011, the Syrian Air Force had begun upgrading their fleet of MiG-29s with assistance from Mikoyan technicians. The plan was to configure them in a manner similar to the modernized MiG-29SM. To achieve this, Mikoyan developed a new avionics package based on the older export-variant Fulcrums previously delivered to Syria. These updated Syrian MiGs featured the more advanced N019ME pulse-doppler radar, improved comms and navigation systems, and a higher payload capacity. These updated jets also gained the ability to launch the R-77 air-to-air missile. Meanwhile, the Syrian Air Force acquired a number of Talisman airborne jamming pods from Belarus.

In September 2015, the Russian Air Force officially entered the conflict. At the behest of al-Assad, a Russian expeditionary force (consisting of a sizeable air arm) arrived in Syria. At first, this aerial task force (known as the Special Mission Air Brigade) arrived in country with an exclusive inventory of Sukhoi aircraft – including Su-30 fighters and Su-25 attack planes. In January 2016, Syrian MiG-29s began flying top cover for the Russian Su-25s on the latter's strike missions against rebel forces. The first of these missions occurred on January 14, 2016 when the Su-25s attacked an ISIS ammunition and fuel depot in Tel-Rifat. The Russian Air Force welcomed this protective cover from the MiG-29s since the "Russian task force and US-led coalition were pursuing different goals in Syria, and the [East-West] relationship was pretty strained."

In September 2017, the Russian Air Force deployed four of its own MiG-29s to the Khmeimim Airbase in western Syria. It would be the first time Russian MiG-29s were deployed in an expeditionary role for extended ground attack missions. In fact, these MiG-29s conducted more than 140 combat sorties against ISIS targets in Syria, along with their secondary duties of escorting the Tupelov Tu-22 bombers.

At first, the Red Star Fulcrums flew only reconnaissance missions. However, they soon graduated into flying strike missions – both as pure MiG formations and alongside a variety of multirole Sukhoi Flankers. A prominent Mikoyan corporate representative, General Designer Sergey Korotkov commented that the MiG-29: "proved itself as a competent fighter and has surpassed the operational requirements – in some respects by a factor of five, and it can be described as an aviation system capable of performing complex tasks in difficult conditions. The mastering of the MiG-29SMT will continue – among other things, with the purpose of checking the efficacy of new advanced aircraft weapons."

Korotkov further stated that the MiG-29's combat performance data would facilitate development of the oncoming MiG-35 and the next generation of Mikoyan fighters. In December 2017, all four MiG-29s returned to their home station at Privolzhskiy Air Base.

Three years later, the Syrian Air Force received its first modernized MiG-29 exports – the vaunted MiG-29SMT variant. As of 2023, the Syrian Air Force had nearly thirty MiG-29s in service. However, on December 8, 2024, following the fall of Damascus to Syrian rebels (and the final collapse of the al-Assad regime), the IDF launched a massive aerial campaign against the former regime's military assets in Syria. At this writing, all available reports indicate that the legacy Syrian Air Force has been destroyed, including most, if not all, of its remaining MiG-29 fleet.

Other engagements

In 2011, at the farthest edge of the Arab Spring, the State of Libya had likewise devolved into a civil war. Following the overthrow and assassination of

Muammar al-Gaddafi, and the temporary peace that ensued, factional violence ignited another Libyan civil war in 2014.

During that second civil war, however, Russia took an active (albeit indirect) role in the conflict. In early 2017, Russia pledged its support for the Tobruk-based rebels rather than the UN-backed Government of National Accord (GNA). In fact, a prominent rebel leader, Field Marshal Khalifa Haftar, had visited Moscow several times throughout 2016 and, in January 2017, he was given a tour of the Russian aircraft carrier *Admiral Kuznetsov*, where he reportedly met with Russian defense officials to discuss weapons contracts.

The ostensible proof of Russian involvement, however, came in May 2020, towards the end of the conflict. Satellite imagery revealed a MiG-29 with Russian markings at the Al Jufrah Air Base in Libya. This sighting fit within a myriad of other reports stating that half a dozen MiG-29s, along with a handful of Su-24 attack planes had landed in Libya. Days earlier, a flight of Russian MiG-29s had been photographed in western Syria. These were (presumably) the same MiG-29s that landed in Libya, ostensibly stopping in Syria to refuel before making their final leg into Al Jufrah.

Later that year, the Director of Intelligence at US Africa Command, Rear Admiral Heidi Berg, confirmed that two Russian-marked MiG-29s had crashed in Libya: one lost on June 28; the other on

September 7, 2020. It is believed that these Red Star MiGs weren't being flown by Russian Air Force pilots, but by mercenary contractors from the infamous Wagner Group. Wagner, a private military company with close ties to the Russian Defense Ministry, has employed several thousand mercenaries, operating in Syria, Sudan, and Ukraine among other hot spots.

More than a dozen Russian-sourced planes were known to have operated in Libya during the Second Libyan Civil War. Although that second conflict formally ended in October 2020, sectarian violence reignited in 2023. And, as before, Wagner Group mercenaries have been seen participating in the conflict, but the extent of their contracted air power (if any) remains unclear.

There has been only one recorded incident of a Cuban MiG-29 firing in anger. On February 24, 1996, three Cessna Skymaster planes registered to the US humanitarian group, *Brothers to the Rescue*, took off from Miami. Formed by Cuban refugees, *Brothers to the Rescue* is a Miami-based nonprofit organization whose mission is to assist waterborne refugees in their efforts to flee the Communist government. Months prior to the shootdown, planes registered to *Brothers to the Rescue* had breached Cuban airspace, dropping leaflets to encourage a mass exodus from Castro's regime.

On the morning of February 24, Cuban Anti-Aircraft Defense detected the three Cessna planes while they were still north of the 24th Parallel, and thus beyond Cuban airspace. US and Cuban reports differ regarding the exact location of the Cessna planes, but all three were intercepted by a two-ship Cuban MiG patrol piloted by twin brothers Lieutenant Colonels Lorenzo and Francisco Perez, respectively flying a MiG-29 and MiG-23. Francisco Perez, aboard the MiG-23, was providing a communication relay for his brother to the Ground Control Intercept station. Lorenzo, in the MiG-29, shot down two of the three Cessnas with his R-60 missiles. The third Cessna, piloted by *Brothers to the Rescue* founder Jose Basulto, escaped and returned to Miami. The incident drew fierce condemnation from the international community, including the UN Security Council. But the Cuban government defended its actions, saying:

> "This is not the case of an innocent civilian airliner that, because of an instrument error, departs from an air corridor and gets into the airspace of another country. These people knew what they were doing. They were warned. They wanted to take certain actions that were clearly intended to destabilize the Cuban government and the US authorities knew about their intentions."

MiG-29s have also seen limited action in Yemen. Following the unification of North and South Yemen in 1990, the newfound Republic of Yemen inherited several MiG-29s from the latter-day South Yemeni forces. The united Yemeni Air Force (YAF) retained its fleet of MiG-29s throughout the 1990s and, in 2001, purchased an additional thirty-six upgraded airframes for delivery the following summer. These upgraded MiGs featured the N019MP radar, making them the most advanced aircraft in the YAF inventory.

Since the start of the Houthi Insurgency and the current Yemeni Civil War, the MiG-29 has seen action on both sides of the conflict. On January 22, 2015, Yemeni President Abdrabbuh Mansur Hadi was forced to resign by the Houthi insurgents and was placed under house arrest. The Houthis named a Revolutionary Committee to assume the powers of the presidency, but Hadi soon escaped house arrest, fleeing to his hometown of Aden. Now beyond the reach of his captors, Hadi rescinded his resignation, denounced the Houthi takeover, and called upon Saudi Arabia for a military intervention.

The Houthis reacted by advancing across multiple fronts and bought the cooperation of some disaffected YAF pilots. Indeed, at the Al-Dailami Air Base, rebels bribed a MiG-29 pilot and two Su-22 pilots to fly combat missions for them. Under these auspices, the Su-22s would bomb the makeshift presidential compound in Aden while the MiG-29 flew top cover.

On March 25, 2015, the now Houthi-controlled MiG-29 and Su-22s went aloft on their first bombing run. Hadi loyalists within the Yemeni Army, however, responded with a heavy barrage of anti-aircraft fire. Although the loyalists failed to down the Houthi-controlled planes, the intensity of their ground fire was enough to dissuade the MiG and Sukhois from returning. During the attack, it was also reported that a flight of pro-Hadi MiG-29s had been scrambled from Al Anad Air Base, to protect the President from Houthi airstrikes.

This defensive scramble, however, was the last known action taken by the YAF.

As the insurgency dragged on, much of the YAF personnel deserted their positions, leaving their aircraft and maintenance hangars unattended. As a result of the Saudi-led airstrikes, most of the YAF's assets were destroyed on the ground throughout 2015-16. These airframe casualties included two MiG-29s destroyed on the ground at Sanaa Airport on April 28, 2015. As of 2023, nearly a dozen Yemeni MiG-29s remain unaccounted for, although they were likely destroyed or parted out for sale on the international arms market.

Sporting the first-generation color scheme of the now-independent, post-Soviet, Ukrainian Air Force, this MiG-29 stands on display at the Abbotsford Air Show in British Columbia, Canada, 1992. *San Diego Air and Space Museum*

The calm before the storm. A Ukrainian MiG-29 Fulcrum takes off from Starokostiantyniv Air Base on October 9, 2018. This flight was part of the "Clear Sky 2018" aerial maneuvers. Clear Sky was a multi-national air exercise involving the Ukrainian Air Force and several NATO members (including the US, UK, Denmark, Estonia, Poland, Romania, and the Netherlands) to promote "regional stability and security, while strengthening partner capabilities and fostering trust." *US Air Force*

Chapter 6:
The Road to Kyiv

When the Soviet Union collapsed in 1991, Ukraine had the highest quantity of Soviet aircraft within its borders – second only to Russia. The newly-independent Ukraine established its own Air Force on March 17, 1992. Taking assets from the former Soviet Air Forces and Air Defense Forces (along with aircraft withdrawn from the former East German *Luftstreitkräfte*), the Ukrainian Air Force received anywhere between 216-280 MiG-29s during the first two years of its independence. Thus, Ukraine had the second-largest fleet of MiG-29s in the world.

However, throughout the 1990s and early 2000s, Ukraine's economic situation could not permit the maintenance of such a large defense apparatus. Consequently, the Ukrainian Air Force began to sell many of its MiG-29s to the other former Soviet states (namely Kazakhstan and Azerbaijan) and beyond. By the late 2010s, these sales had left Ukraine with fewer than 50 operational MiG-29s, stationed among a handful of airbases within the regional air commands. Following the onset of the Ukrainian Civil War (involving separatists in Donbass and the Russian annexation of Crimea) and the subsequent Russo-Ukrainian War, the Mikoyan Design Bureau – now under the banner of the United Aircraft Corporation

– withdrew its technical support for all Ukrainian MiGs, forcing Ukraine to rely on its own resources to support the aging fleet of MiG-29s.

It was under the auspices of Ukraine's Civil War that the Ukrainian MiG-29s had their first test in combat. The political crisis began in 2013 with the overthrow of President Viktor Yanukovich. A coalition of far-right wing nationalist groups filled the ensuing power vacuum, which further accelerated tensions between the western and eastern parts of the country. The former region was aligned with Western Europe, while the latter was decidedly pro-Russian and mostly Russian-speaking. The eastern region of Ukraine refused to recognize the new nationalist Kyiv government, which first led to the Crimean crisis, followed by the Donbass independence movements. Within the Donbass region, two self-proclaimed "republics" asserted their independence from Ukraine: the Donetsk People's Republic (DPR) and the Luhansk People's Republic (LPR).

In April 2014, the Ukrainian government responded by sending military units into the disputed region. Ukrainian ground forces led the way, assisted by a few Air Force attack units. The DPR and LPR, meanwhile, had their own militias, supplied with whatever they could commandeer from local police stations and military bases.

Ukrainian Air Force Su-24 and Su-25 attack planes carried most of the burden for the air campaign, but a few MiG-29s from the 40th and 114th Tactical Aviation Brigades were also committed to the fight. At first, the MiG-29s flew "presence patrols" and show-of-force missions over DPR and LPR strongholds, hoping to affect a psychological impact on the local militias. Later that summer, however, the Fulcrums began flying strike missions over enemy territory.

The results, however, were disastrous for the Ukrainian Air Force.

Indeed, within ten days, two MiG-29s had been shot down by enemy ground fire. On August 7, 2014, a MiG-29 piloted by Colonel Yuliy Mamchur was hit by a MANPADS rocket near Zhdanovka in the Donetsk People's Republic. Mamchur safely ejected from the aircraft (and was later elected to Ukraine's parliament) but his ill-fated MiG crashed near Rozovka, leaving nothing left of the airframe. Ten days later, the LPR militia downed another MiG-29 from the 40th Tactical Aviation Brigade with SAM fire in the Krasnodon District. Like Mamchur, the pilot of this ill-fated MiG ejected to safety, but his plane was destroyed on impact with the ground. Following these double kills, Ukraine withdrew its MiG-29s from the warzone.

The respite, however, was short-lived. For on February 24, 2022, Russian forces launched a full-scale invasion of Ukraine. Vladimir Putin, the Russian President, had long harbored ideations to annex

Ukraine. Whatever the reasons for his decision, he expected Russian air and ground forces to make short work of the Ukrainian resistance. Military experts and media analysts predicted a complete collapse of Ukrainian defenses within a few weeks to a few months.

But even the most seasoned analysts failed to anticipate the ferocity of Ukraine's resistance.

At this writing, the war in Ukraine is still ongoing; and both sides have claimed the upper hand during any given season. But wherever the cards may fall, there can be little argument that the MiG-29 has played a prominent role in the air war over Ukraine.

The first wave of the air campaign began at 5:00 AM (local time) on February 24. A total of 75 Russian aircraft were committed to the initial onslaught, accompanied by a fusillade of land-based ballistic and naval-based cruise missiles. Early battle damage assessments confirmed more than 80 Ukrainian targets had been destroyed, including eleven airfields. A near-concurrent cyber-attack, meanwhile, crippled many of Ukraine's Command & Control nodes.

The Ukrainians, though shaken by the multi-pronged attack, fought back valiantly. At the Ivano-Frankivsk Air Base, for example, MiG-29 pilots from the 40th Tactical Aviation Brigade manned their aircraft at just after 3:00 AM, but were held on the runway for nearly two hours before the first ready-alert MiGs were ordered to take off. The timing was fortuitous: mere seconds after the MiGs scrambled from the airfield, the pilots saw gigantic flashes in their rear-view displays – the telltale sign of Russian missiles impacting on the runway at Ivano-Frankivsk.

Following this initial bombardment, the Russians quickly turned their attention towards Kyiv. Now that the early warning radar systems in the Ukrainian capital were neutralized, Russian AWACS directed several pairs of Su-30 and Su-35 Flanker interceptors to occupy combat air patrol (CAP) stations north of Kyiv. Meanwhile, despite the initial shock and inertia of mobilizing its forces under duress, the Ukrainian Air Force scrambled its MiG-29s from the 40th Brigade to meet the incoming bandits.

According to Ukrainian sources, the lead pair of MiG-29s entered Kyiv airspace and were detected almost immediately by a flight of Su-35s. The lead MiG-29, piloted by Colonel Vyacheslav Yerko, successfully downed an Su-25 Frogfoot and two Mi-24 helicopters, before he himself was shot down by an R-77 missile from one of the intercepting Su-35 Flankers. The charred remains of Yerko's plane were found in a forest in the Sosnivka region, southwest of Kyiv.

Shortly thereafter, another pair of MiG-29s scrambled from Vasilkyiv Airbase, whereupon they encountered another flight of Russian Su-35 Flankers. Sadly, the Ukrainian Fulcrum piloted by Lieutenant Vyacheslav Radionov was downed by an R-77 missile. The other MiG-29 was seen on video firing its S-8 rockets, but it remains unclear whether these unguided rockets hit any Russian targets. Later that day, another MiG-29 from the 40th Tactical Aviation Brigade, piloted by Lieutenant Roman Pasulko, was shot down near Vsyhgorodsky in northern Kyiv. Pasulko ejected from the stricken MiG-29, but sadly did not survive the ejection.

All told, the greatest threat to Ukrainian MiGs on the first day of the air war came not from enemy fighters, but from Russian electronic warfare. Indeed, published accounts from Ukrainian MiG-29 pilots confirmed that their onboard radars were repeatedly jammed by Russian ECMs. These same MiG-29s tried to engage with their R-27 missiles, but the superior Russian Sukhois had them outgunned with their R-77s. According to famed aviation analyst Tom Cooper: "The Russians enjoyed the advantage of their missiles having active radar homing in the terminal flight phase." This meant that the R-77 was more of a "fire-and-forget" system – the Russian pilots did not have to track or guide the missile to its target. "Moreover, the high-flying Russians could launch their missiles from high altitudes and at high speeds, thus extending their range."

By the second day of the air war, however, an unexpected "hero" had emerged from the ranks of Ukrainian MiG-29s. In the airspace over Kyiv, a lone MiG-29 pilot was purportedly blasting Russian jets from the sky with impunity.

Locals were calling him the "Ghost of Kyiv."
Within a day and a half, Ukrainian media outlets were reporting that the Ghost of Kyiv had downed at least six enemy fighters, including: one Su-27; two Su-35s; one Russian-flagged MiG-29; and an additional Su-25 attack plane.

Downing six aircraft alone would make him an "ace" - the typical standard is five kills.

But achieving that status in one day?

That would make him the first man in Europe since World War II to earn the title "ace in a day." And thus far, he would be the only ace of the 21st Century.

Given the sensational nature of the story, it wasn't long before the media went ballistic. News outlets worldwide began circulating photos and video footage claiming to show the Ghost of Kyiv. Many of these early clips were proven to be doctored video game footage, but Ukrainian social media soon began posting photos of a pilot whom many identified as the Ghost of Kyiv. For example, former Ukrainian president Petro Poroshenko posted a photograph to Twitter (now "X") of a fighter pilot, claiming it to be the Ghost of Kyiv. However, the photo was later found to be an unrelated image of Captain Andrii Pilshchykov, another MiG-29 pilot, but certainly not the Ghost of Kyiv. This and other photos, too, were quickly discredited.

A few days later, on February 27, the Security Service of Ukraine made a Facebook post claiming that the Ghost of Kyiv had now destroyed *ten* enemy aircraft.

But who was he?

And where were these numbers coming from?

By now, the mainstream print media had jumped onboard. On March 3, the *Times of London* reported that a Ukrainian source had confirmed the Ghost of Kyiv was real and that he was still alive.

Then came a report that the "Ghost of Kyiv" was Major Stepan Tarabalka, a 29-year-old MiG pilot who had been shot down during a dogfight near Zhytomyr in northern Ukraine. Tarabalka was a valiant pilot, but he wasn't the Ghost of Kyiv.

Still, as the legend grew, so too did the stories surrounding it. At one point, some sources claimed that the Ghost had downed as many as 40 enemy bandits.

But while the world was trying to figure out whether this Ghost was real, the artist community got straight to work, producing lovely renditions of who (or what) this Ghost may have been. And it wasn't long before the internet memes started popping up as well.

However, on April 30, 2022, the Ukrainian Air Force confirmed that the Ghost of Kyiv was only a myth. Saying:

> "The ghost of Kyiv is a superhero-legend, whose character was created by Ukrainians," [but] "the #GhostOfKyiv is alive, it embodies the collective

spirit of the highly qualified pilots."

Many netizens were openly disappointed by the news, including those who doubted the story's veracity from the beginning. Given the logistical and sortie generation requirements for a MiG-29, it didn't seem likely that a single Ukrainian Fulcrum could down six planes in just one day. It wasn't impossible, just highly unlikely.

Two years later, in early 2024, Ukrainian defense spokesperson Illia Yevlash, admitted that the Ghost of Kyiv was a fictional hero created by his public affairs team after the Ukrainian Air Force began tallying up aerial victories against Russian planes. "During a brainstorming session," Yevlash recalled, "Volodymyr Fityo [a public affairs specialist for the Ukrainian Ground Forces] suggested calling him the 'Ghost of Kyiv.' Everyone supported the idea. Later, the news agencies spread the info. We wrote a post about him twice, and then the 'Ghost' began to live on his own."

An anonymous Ukrainian military expert then told BBC News that the Ghost of Kyiv helped "raise morale at a time when people need simple stories," but that the Ukrainian Air Force still warned people not to "neglect the basic rules of information hygiene" and to "check the sources of information before spreading it."

Still, this wasn't the first time that a mythical pilot had been created as a part of wartime propaganda.

For example, the Vietnam People's Air Force created the infamous Colonel Toon, the mythical North Vietnamese flying ace. And there are many indications to suggest that Captain Mohammad Rayyan, the famed "Sky Falcon" of the Iran-Iraq War, was also a myth.

But although the Ghost of Kyiv was merely a legend, he nevertheless had a tremendous impact on the morale of Ukraine's fighting forces. It became a case study in the broader realm of psychological warfare, disinformation, and "black propaganda" as the latter-day OSS would call it. Black propaganda is designed to rally support for the friendly cause while simultaneously getting inside the enemy's head, wearing down his confidence, and weakening his resolve. Whatever the outcome of this war may be, historians and wartime analysts will undoubtedly point to the Ghost of Kyiv as a true testament to the power of wartime mythology.

As February turned to March, the situation continued to look bleak for the Ukrainian MiG-29s. On February 24-25, for example, two MiG-29s were confirmed to have crashed; one near Kharkiv, the other near Ivano-Frankivsk. Neither was attributed to enemy fire, but the following day, a Russian missile strike destroyed at least six MiG-29s on the ground at Ivano-Frankivsk International Airport.

On the night of March 1, however, Ukrainian MiG-29s claimed an aerial victory against a flight of

Russian Su-35 Flanker-M fighters. According to the Ukrainian Air Force Command: "a fierce air battle broke out in the Kyiv region between a pair of MiG-29 fighters...and a pair of Russian Su-35 fighters. As a result of the air battle, both Russian aircraft were destroyed." The official press release did not indicate the unit to which the victorious Fulcrums belonged, but it was later confirmed to be the 40th Tactical Aviation Brigade. Ukrainian officials went on to say that the MiG-29s were assisted in their efforts by the local SA-10 SAM missile batteries. Unfortunately, during this aerial skirmish, the MiG-29 piloted by Oleksandr Brynzhala was shot down by one of the Russian Flankers. On April 2, Oleksandr Brynzhala was posthumously named "Hero of Ukraine," and honored with the Order of the Gold Star.

Brynzhala's wingman, a pilot identified only as "Volodymyr," gave a slightly different rendition of the encounter when interviewed by Ukrainian defense correspondent Anastasia Olekhnovych. Although he confirmed that Oleksandr Brynzhala was a hero who shot down two enemy planes, Volodymyr recalls that they were engaged against *twelve* enemy fighters, not two, as the Air Force had initially stated. "We thought there were only four of them," he said. "But they attacked from several directions at the same time with 12 planes! Everything happened instantly. In close combat, our chances are somewhat equal, although the quantitative and technological advantage is still on their side. [Brynzhala] shot down two enemy planes, I shot down one. The maneuvers he performed, his determination and accuracy are beyond the power of any occupier!"

Volodymyr also stated that, of late, the Russian pilots had changed their tactics, opting to avoid close-quarter dogfights and the envelope of Ukrainian air defenses. Instead, Volodymyr noted that the enemy was devolving into hit-and-run tactics, launching missiles from maximum standoff ranges, then running away.

But although the Russians seemed to be avoiding toe-to-toe engagements, Ukrainian MiG-29 losses continued to mount. Indeed, by the end of 2022, Ukraine had lost twenty-seven Fulcrums to enemy fire.

These losses included the March 23 shootdown of Lieutenant Colonel Dmytro Chumachenko from the 204th Tactical Aviation Brigade. Ukrainian sources reported that Chumachenko was shot down by an enemy fighter (possibly an Su-27) over the Zhytomyr Oblast. Very little is known about Chumachenko's engagement, but forensic analyses concluded that he had steered his ailing MiG away from the nearby homesteads, opting to crash into the local forest belt rather than endangering civilians on the ground. Chumachenko's story was even more astounding considering that he was born in Russia and had immigrated to Ukraine years earlier, becoming a proud Ukrainian citizen.

Sadly, in the immediate aftermath of his death, the

wreckage of Chumachenko's MiG became a target for nearby looters and scavengers. Within minutes of the impact, five residents from the nearby village of Denishy descended onto the stricken MiG. Without any regard for the dignity of Chumachenko's body, these opportunistic scavengers stole ammunition, avionics, and microcircuits with precious metals, as well as Chumachenko's sidearm.

Luckily, the matter was reported to a local precinct of the Ukrainian national police. By early May, the suspects had been identified, and a police raid recovered the items taken from the crash site. Nevertheless, the story was a stark reminder that during times of war, black-market profiteers can emerge on any side of a conflict, with their loyalties belonging only to the highest bidder.

On September 25, a MiG-29 piloted by Major Taras Redkin was downed by enemy fire during a Suppression of Enemy Air Defense (SEAD) mission over Bashtanka Raion. According to the Ministry of Defense: "Taras was performing a task to destroy enemy air defenses...not far from the line of combat contact. As usual, he had effectively worked on ground targets and was already turning around." At about 4:00 AM local time, his last words on the radio were: "I see launches towards me," before his plane disappeared from radar. Although international news outlets claimed he was downed by a Russian fighter, it is more likely that Redkin was killed by an enemy SAM, given the nature of SEAD operations and the terminology he used during his radio transmission.

Two weeks later, on October 12, a MiG-29 piloted by Major Vadym Voroshylov was lost while intercepting a flight of Russian Shahed-136 (Geran-2) drones over Vinnytsia. Considering their function, these Iranian-built drones were among the most notorious aerial weapons of the conflict. By design, they were "loitering munitions" – meaning they would loiter until finding a target to crash into, detonating upon impact. More commonly, these unmanned systems were known as "suicide drones" or "kamikaze drones." Voroshylov had already downed three such drones earlier in the day; and on this current mission, he successfully downed two more. But the explosion from the second drone riddled the cockpit of Voroshylov's MiG, forcing him to eject. Luckily, Voroshylov survived the ejection (in fact, he posted a selfie on social media later that day, highlighting his blood-soaked, battle-weary visage) while the stricken MiG crashed without any collateral damage to the local civilians.

For his actions on October 12, Voroshylov was named a "Hero of the Ukraine." Downing five enemy drones earned him the distinction of being an "ace" – but five aerial victories within twenty-four hours put him in the distinguished class of aviators known as "ace in a day." His combat record made him the first European pilot since World War II to hold that distinction. Ironically, Voroshylov became the real-life manifestation of the bygone Ghost of Kyiv.

Meanwhile, Ukraine had been soliciting NATO for any military aid it could render. Poland and Slovakia (two former allies from the Eastern Bloc era) pledged to donate their excess MiG-29s, but the United States made a surprising delivery of AGM-88 HARM missiles to Ukraine in late 2022.

Although it had a reputation for being one of the best air-to-ground missiles in the world, the AGM-88's arrival was perplexing because the system wasn't compatible with any Soviet-built aircraft. Indeed, none of the MiG or Sukhoi wing pylons could accommodate the missile fitment; and the AGM-88 needed a digital interface to operate.

Nevertheless, Ukrainian aerospace engineers and American contractors developed a solution, adapting the NATO-spec missile to the MiG-29's airframe. They installed specially-modified wing pylons and equipped the pilot with an iPad (and similar touchscreen tablets) in the cockpit to operate the air-to-ground missiles.

US Undersecretary of Defense for Acquisition and Sustainment, Dr. William LaPlante, confirmed in 2024 that the Ukrainian Air Force had implemented these adaptive measures. "Working with the Ukrainians," he said, "we've been able to take many Western weapons and get them to work on their aircraft where it's basically controlled by an iPad by the pilot. And they're flying it in [the] conflict like a week after we get it to him." After integrating the AGM-88 HARM,

Ukrainian MiG-29s also began using the Joint Direct Attack Munition-Extended Range (JDAM-ER) precision-guided bombs and the French-built Hammer AGM.

But even with the arrival of these new air-to-ground munitions, the MiG-29 was still at a disadvantage given the limitations of its organic fire controls. Because the Fulcrum could not accommodate the AN/ASQ-213 targeting system, the pilot had to fly much closer to the anticipated ground target before launching his missile. By contrast, the AN/ASQ-213 would allow the pilot to track the location of hostile radar systems and engage targets from greater standoff distances, thereby minimizing exposure to enemy air defense batteries. In the aftermath of Major Redkin's MiG-29 shootdown in 2022, Ukrainian defense analysts said the incident demonstrated Ukraine's need to acquire the F-16 as a means to facilitate better SEAD missions, since the F-16 came with the AN/ASQ-213. The remark was surprisingly prescient; for in 2024, Ukraine took delivery of its first F-16s donated from Belgium, Denmark, and the Netherlands.

Throughout the Spring of 2023, Ukrainian MiG-29s continued flying sorties in the face of diminishing supplies, fierce opposition, and the ever-present fog of war.

On January 7, for example, Ukrainian air defense troops shot down one of their own MiG-29s near

Donetsk. According to various sources, the perpetrator was an SA-8 "Osa" missile launcher, while the ill-fated MiG belonged to the 114th Tactical Aviation Brigade. Tragic as it was, this friendly fire incident was far from unprecedented; Ukrainian forces had previously downed a friendly Su-27 over Kyiv in the opening days of the war.

Meanwhile, the suicide drones continued wreaking havoc over Ukrainian airspace. One such drone (another dreaded "Geran-2") somehow precipitated the downing of a MiG-29 piloted by Lieutenant Dmytro Shkarevsky. He intercepted the loitering munition, skillfully downing it with a missile shot.

But moments later, his own plane began to break apart.
Bewildered, Shkarevsky ejected from the stricken aircraft, suffering serious wounds from the falling debris and his parachute landing in the Samara River.

Still unsure of who (or what) had shot him down, Shkarevsky pulled out his cell phone to contact his commanding officer. After confirming his location, Shkarevsky's commander dispatched a rescue team with an estimated arrival time of forty minutes. In the meantime, Shkarevsky sought shelter with a local family until he was recovered by friendly forces and treated for his wounds, including a broken vertebra and numerous skin lacerations from the aircraft shrapnel.

It was unlikely that Shkarevsky's MiG was shot

down by an enemy bandit. There were no Russian fighters reported in the area; and the Russian Ministry of Defense hadn't claimed downing a MiG-29 on that date. Reflecting on the crash, however, Shkarevsky was convinced that his MiG had been the victim of mechanical failures and subpar maintenance. He publicly commented on the qualitative disparity within Ukraine's MiG-29 fleet - some aircraft were in much better shape than others, and maintenance standards were not equally enforced across all units.

That summer, Ukraine lost another MiG-29, piloted by Captain Vladyslav Saveliev from the 114th Tactical Aviation Brigade. On the morning of June 2, 2023, Saveliev was flying his eighth combat mission over Donetsk, during which he destroyed a mobile enemy headquarters with his adapted AGM-88. However, while returning to base, Saveliev's plane was hit by a Russian SAM.

Coincidentally, one of Saveliev's squadron mates was Captain Andrii Pilshchykov, the man whose portrait had been widely circulated as the Ghost of Kyiv during the early days of the war. Pilshchykov, although not airborne during the mission, claimed to have been monitoring the radio traffic while on the ground. According to Pilshchykov, Saveliev executed the mission perfectly, and tried to shake off the SAM to the greatest extent that the MiG-29 would allow. "Vlad died instantly due to a powerful explosion," he said. "He died with dignity, in battle [and] satisfied with the successfully-completed job." The following day, June 3, Pilshchykov avenged his friend's death – destroying a Russian air defense site in the same sector where Saveliev had fallen.

Interestingly, a year prior to the war, Saveliev had enrolled in a US-sponsored training program at Columbus Air Force Base, learning advanced fighter tactics and maneuvers. When the war began in 2022, he was understandably eager to return to Ukraine, but had to wait an additional twelve months before completing his coursework. He returned to Ukraine in March 2023, nearly three months to the day before his untimely death.

Back at the Ivano-Frankivsk airstrip, Saveliev's comrades honored his memory via an Anglophone tradition that Saveliev had learned while stationed in America: *the ceremonial burning of a piano.* Indeed, the RAF and US Air Force had a long-standing tradition of setting a piano on fire as a means to honor their fallen comrades. The sacrificial piano for Saveliev's ceremony bore the word "Nomad" painted on the front – a nod to the callsign Saveliev had been given by his American counterparts. On the side of the piano appeared a prominent number "12" written in white paint, representing the "White 12" designation of the individual MiG.

The rest of the summer was relatively quiet along the aerial front. By Fall 2023, however, the mood was decidedly different. On September 19, for example, a Ukrainian MiG-29 sitting on the tarmac at Krivoi Rog

Air Base became the latest victim of a Russian suicide drone. This time, however, the perpetrator was not the ubiquitous Geran-2 import drone, but a Russian-built ZALA Lancet making its combat debut. The new loitering munition had found its way over the air base, where it simultaneously destroyed two Su-25s alongside the MiG-29 during the same attack. More alarming than the attack itself was that a *second* drone had provided overhead cover, observing the Lancet whilst it initiated the attack. Two enemy drones breaching the Krivoi Rog airspace was a clear indication that local air defenses were either inactive or weren't working properly.

By late October, reports had surfaced claiming that Russian forces had downed *seventeen* MiG-29s, including seven destroyed within a single day. However, these figures have not been independently verified, and Ukrainian authorities have yet to confirm these losses. Meanwhile, Ukrainian forces claimed that one of their MiG-29s had successfully downed a Russian drone over Zaporizhzhia, but did not specify the type of drone involved.

In November, however, the Ukrainian Air Force did confirm that an unidentified MiG-29 pilot was shot down by a Russian SA-21 Growler missile during a SEAD mission over Donetsk. The pilot ejected and survived with minimal injuries. Although the Ministry of Defense released nothing more than the pilot's callsign, they confirmed that he had previously flown more than 150 combat missions, attacking enemy air defenses with the latest imported HARM missiles and JDAMs. Other sources claim that he downed as many as ten Russian drones within the past year.

Throughout the latter half of 2023, Ukrainian forces mounted an air-ground counteroffensive designed to recapture the occupied territories in Donetsk and Zaporizhzhia. Although Ukraine successfully recaptured fourteen villages in the Donetsk-Zaporizhzhia area, the campaign was considered a negligible victory; and by some accounts, it was even called a "stalemate." Eastern military analysts were even less forgiving: saying that Ukraine's failure to re-capture key cities or reach the Sea of Azov was tantamount to a defeat.

Meanwhile, on the aerial front, Ukrainian MiG-29s continued standing strong in the face of enemy fire. According to most estimates, the year-end losses for Ukrainian Fulcrums totaled between 9-13 (with several more unconfirmed losses reported in various news outlets). Although Ukrainian MiG sorties continued to make inroads against enemy air defenses, the Russians' anti-air capabilities were at least effective enough to disrupt Ukrainian air support to the frontlines.

By Spring 2024, although the Ukrainian offensive had stalled on the ground, there was still plenty of action in the skies. On January 5, 2024, a Ukrainian MiG-29GT (the variant donated from Poland) piloted by Vladyslav "Blue Helmet" Zalistovskyi of the 114th

Tactical Aviation Brigade crashed while on a mission near Luganka in the Kirovograd Oblast. Initial reports suggested that Zalistovskyi was downed by enemy fire; but mechanical failure soon emerged as the most likely culprit.

The Russian Ministry of Defense, known for its pride in reporting downed Ukrainian jets (whether truthfully or not), didn't claim responsibility for Zalistovskyi's MiG. In fact, there was no mention of a MiG-29 in the defense briefings for January 5. While some have speculated that Zalistovskyi may have been the victim of fratricide from Ukrainian air defenses (or that the crash was due to pilot error), mechanical failure remains the most viable cause. Indeed, the MiG-29GT variant (which Poland had received from the German *Luftwaffe* years earlier) had been a perennial cause for concern in the Polish defense community. Government ministers and military analysts had publicly stated their concern for the poor reliability and maintenance procedures of the Polish MiG-29 fleet. These public outcries came following a number of high-profile accidents resulting from mechanical failures. Polish parliamentarian Maciej Kopiec said of the latter-day MiG-29s:

> "Problems with the technical condition of machines appear without interruption. According to them, engineers struggle, among others with cabin malfunctions, dissection of fairings and cracking chassis components."

Yet despite these misgivings, Poland nonetheless

donated its MiGs to the Ukrainian cause. Tragically, Vladyslav Zalistovskyi did not survive the crash. He was 23 years old.

Sadder still, Zalistovski wouldn't be the last casualty aboard an ill-fated Polish MiG-29. Indeed, on April 27, 2024, Colonel Valentin Korenchuk, one of the many pilots who had been credited as the "Ghost of Kyiv," died on a combat mission while piloting a Polish-import MiG-29A. As of yet, Ukrainian sources have not publicly revealed the cause of Korenchuk's death, but given the poor condition of these donated MiGs, Korenchuk, too, may have been another victim of mechanical failure. The Russian Ministry of Defense did, however, issue a press release claiming to have brought down a MiG-29 during the week of April 27. Given the scarcity of corroborative sources, however, the true cause of Korenchuk's death remains unknown.

Nevertheless, Korenchuk was a highly-regarded pilot, and many considered him to be the best fighter pilot in Ukraine. At the time of his death, he was commanding a squadron of the 40th Tactical Aviation Brigade, and had flown more than 80 combat missions. It was also revealed that Korenchuk had been Colonel Vyacheslav Yerko's wingman on the mission of February 24, 2022, wherein Yerko downed a Russian Su-25 Frogfoot and two enemy helicopters before falling to a Su-35 "Flanker" the same day.

By the summer of 2024, the Russo-Ukrainian War had

taken an interesting turn. In May of that year, Russian forces began a renewed offensive in the Kharkiv region along the Russia-Ukraine border. Their stated goal was to create a "buffer zone" for the beleaguered border regions. Despite some early advances, Russian forces soon lost their momentum in the face of Ukrainian counterfire. That August, Ukraine launched its first offensive into Russian territory, an audacious cross-border advance into the neighboring Kursk Oblast. Ukrainian forces took advantage of the inexperienced units defending the borderlands along Kursk, and they were able to seize a considerable swath of territory during the opening days of the campaign. The downstream benefit to Ukraine's military was that the cross-border incursion prompted Russia to divert thousands of troops from occupied Ukrainian territory to counter the threat in the Kursk borderlands.

Ahead of this incursion, however, Ukrainian MiG-29s scored their first victories against ground targets inside Russian territory. In June 2024, a flight of Ukrainian MiG-29s struck a regimental command post at Nekhoteyevka in the Belgorod Oblast (about one kilometer from the Russia-Ukraine border). Satellite imagery confirmed destruction of the Russian command post; and forensic data indicated that the weapon of choice was a NATO-spec missile, possibly a French-built Hammer. The Ukrainian General Staff didn't comment on the type of missile used, but confirmed that the target had been the command post of a Russian motorized rifle regiment.

But as the Kursk Offensive got underway, Ukraine lost another MiG-29 to enemy fire. On August 12, Captain Olexander Migulya, a 27-year-old Fulcrum pilot from the 40th Brigade was shot down over Donetsk. However, the circumstances behind his downing remain unclear. Ukrainian civil documents confirm that his MiG crashed at approximately 8:40 AM in the vicinity of Rozkishne in Donetsk. Some sources claim that he was downed by a Russian Su-30, while others have claimed that the perpetrator was a MiG-31 or an SA-21 missile. However, there has been no viable evidence to validate these theories.

As had been the case with Vladyslav Zalistovskyi, the Russian Defense Ministry did not claim any MiG-29 kills for the date in question. Images of the crash site confirmed that the plane was Migulya's and, considering his armaments for the day's mission, it's likely that his MiG crashed while attempting to fire one of his air-to-ground missiles. This would explain why Migulya's MiG wasn't mentioned in the Russian daily briefings for August 12.

Ironically, one year prior to the crash, Migulya was interviewed by the Ukrainian Air Force for a video published in late 2023. Discussing the trials and tribulations of Ukrainian MiG pilots, Migulya commented on the recent arrival of the F-16 in Ukraine, stating that Ukraine's current fleet of fourth-gen Eastern Bloc fighters were both "physically and doctrinally outdated," leaving their pilots vulnerable to Russian air superiority.

By 2025, the Ukrainian-led Kursk Offensive had stalled, and Russian Ground Forces had begun to reclaim the occupied territories. Simultaneously, the MiG-29 nearly became a victim of inter-service cannibalizing. In 2024-25, amidst the growing need to replenish troops on the frontlines in Eastern Ukraine, some units began pulling MiG-29 technicians from their airfields, sending them into the Ground Forces to serve as infantrymen.

Throughout 2024, an estimated 2,500 to 5,000 Ukrainian airmen were forcibly reassigned into ground combat roles. Understandably, the decision sparked controversy within the Ukrainian Air Force. Some claimed that this personnel shuffle could reduce the operational readiness of certain air units by up to 40%. Given the already-reduced capacity of Ukraine's air maintenance program, such concerns were not unfounded.

In January 2025, the Ukrainian Air Force established an investigative commission to prevent any further MiG-29 maintenance crews from being involuntarily transferred, and to reverse any reluctant transferees who had already been taken from their hangar bays. According to the Ukrainian Air Force Chief of Communications, Colonel Yuriy Ignat:

> "The commission is working to prevent the transfer of military personnel with critical

specialties, which could significantly impact the combat capabilities of Air Force units. An Air Force commission is currently investigating whether violations have occurred. If such cases are found, corrective actions will be taken."

This statement came on the heels of a video published by Ukrainian Air Force Sergeant Vitaliy Gorzhevsky of the 114th Tactical Aviation Brigade. In the accompanying statement, Gorzhevsky wrote: "We, the aviation technical personnel of the Air Force of the Armed Forces of Ukraine (AFU), have been fulfilling our critical mission on MiG-29 fighters for over a decade. Our unit received a telegram ordering the transfer of nearly all technical staff to the infantry. This will leave us without the personnel needed to maintain our aircraft."

Gorzhevsky further added that 250 maintenance specialists from his unit had already been reassigned; and there were plans to transfer an additional 218 airmen. "The technical staff is essentially being dismantled," he continued, "and without us, aviation cannot function. We are struggling to maintain combat readiness, but without service personnel, it is simply impossible."

His statements were echoed by veteran MiG pilots, such as Vadym Voroshylov, who emphasized that the technical crews and mechanics often have five or more years of advanced military schooling to gain the requisite experience for aircraft maintenance. Thus, sending them to the frontlines would be a waste of human capital, especially for those who were trained in the West.

At this writing (early Spring 2025), the personnel matter remains under investigation, but the Ukrainian General Staff has repeatedly denied any allegations of human resource mismanagement. Rather, they claim instead that any Air Force personnel transferred to the infantry were not "mechanics" or "technicians," but "newly-trained soldiers prior to deployment." But while the matter is still being adjudicated, Colonel Ignat has disclosed that the investigation will not focus solely on finding and retrieving MiG-29 specialists, but also air defense maintenance crews, radio operators, and electronic warfare specialists.

As the War in Ukraine settles into its third year, the MiG-29 is still a featured player in the aerial campaigns against Russian dominance. Although Ukraine has lost most of its antebellum MiG-29 fleet (and the plane is demonstrably being phased out in favor of the F-16), the surviving aircraft and their imported counterparts regularly patrol the war-torn skies, engaging targets of opportunity in the air and on the ground.

Most of these latter-day engagements have been directed towards enemy air defenses or the seemingly-inexhaustible supply of kamikaze drones. In the meantime, reports will continue surfacing on both sides of the conflict, each claiming the destruction of this aircraft or that aircraft, followed by the normal variety of fact-checkers and "independent verifications" to ensure informational hygiene.

Whatever role the MiG-29 plays towards the resolution of this conflict, there can be little doubt that the Russo-Ukrainian War has become the defining event for the Fulcrum in the 21st Century. For Ukraine, the MiG-29 wasn't just a tactical fighter; it became a symbol of resistance – proving that an aging aircraft could hold its own when paired with modern tactics and Western technology.

The war has also seen the MiG-29 operate in ways its original Mikoyan designers never anticipated. Ukrainian engineers and NATO-based contractors modified the aircraft to utilize a variety of Western-built munitions, extending the MiG-29's relevance far beyond expectations.

Whether engaged in air-to-air combat or ground attack missions, many of the pilots became latter-day heroes. Men like Valentyn Korenchuk and Oleksandr Migulya gave their lives, often flying combat missions where they knew the odds of survival were virtually nonexistent. Despite suffering heavy losses from enemy fighters or ground-based air defenses, the MiG-29 has carried on, showcasing the determination of Ukraine's Air Force.

As the war continues, the MiG-29 remains a part of the fight, but its frontline service is drawing to a close. With Ukraine receiving Western fighter jets like the F-16 and Dassault Mirage, the aging Fulcrums are gradually being replaced.

Yet, their legacy in the conflict is undeniable.
The MiG-29 has proven its worth, not just as a Soviet relic, but as a resilient platform capable of adapting to modern warfare. It was the first aircraft to challenge Russian air superiority; the first aircraft adapted to carry NATO-spec weaponry; and the last Soviet-era fighter to stand tall in the face of modern-day threats.

The story of the MiG-29 in the Russia-Ukraine War is one of resilience, ingenuity, and courage. The Fulcrum may not be the most advanced fighter in the skies, but in the hands of Ukrainian pilots, it has become a symbol of solidarity, proving that "air power" isn't just about the planes, but the people who fly them.

The ceremonial piano burning in honor of Captain Vladyslav Saveliev, callsign "Nomad," who piloted the MiG-29 christened "White 12."

Epilogue:
The Enduring Legacy of the MiG-29

For more than four decades, the Mikoyan MiG-29 has left its mark on the history of aerial warfare. Designed during the height of the Cold War, it was a fighter born from the fierce technological rivalry between NATO and the Warsaw Pact. Over the years, it has flown under the flags of various air forces, proving itself to be a formidable – though not always victorious – combat aircraft. From the skies over Iraq and Serbia to the fierce battles in Ukraine, the MiG-29 has nevertheless demonstrated its resilience, adaptability, and an enduring relevance to modern air power.

Initially conceived as an agile air-superiority fighter to counter the American F-15, the MiG-29 was a marvel of Soviet engineering. Compared to other eastern fighters of its day, the MiG-29 had enviable top speeds, a high thrust-to-weight ratio, and exceptional maneuverability. Yet, the collapse of the Soviet Union left its future uncertain.

As Russia and the former Eastern Bloc states struggled to modernize their air forces, the MiG-29 found new life on the export market. From Poland to India, many of the new MiG recipients modified the plane to suit their operational needs. Some were able to bring the MiG-29 into the digital age, while others struggled with maintenance issues – a longstanding drawback of the aircraft's complex design and cyclic maintenance needs. Nevertheless, the Fulcrum remained relevant, often deployed to contested airspace, where it could be a powerful asset if employed and maintained properly.

Throughout the 1990s and early 21st Century, the MiG-29 was tested in a variety of high-stakes combat environments. In the Middle East, for example, MiG-29s were used by the Iraqi and Syrian Air Forces, though their impact was often limited by a lack of modernization, training, and logistical support. In 1999, Serbian MiG-29s squared off against NATO fighters including the F-15 and F-16. Outmatched in terms of avionics and comparative armaments, Serbian MiGs suffered heavy losses but still demonstrated the aircraft's resilience under extreme conditions.

But nowhere has the MiG-29's combat history been more pronounced as in the ongoing War in Ukraine. Despite being outnumbered and outgunned, Ukrainian pilots have used their Fulcrums to great effect – flying low-altitude strikes; intercepting cruise missiles and suicide drones; and engaging Russian fighters in aerial combat. The introduction of NATO weapons such as AGM-88, JDAMs, and other precision-guided munitions, have extended the MiG-29's strike capabilities far beyond what its Soviet designers could have envisioned.

In many ways, the War in Ukraine has reaffirmed

the MiG-29's adaptability, proving that even a fourth-generation airframe, when paired with modern weapons and innovative tactics, can hold its own against a fifth-generation threat.

Stories of Ukrainian pilots flying into combat against overwhelming odds (with many never returning) have cemented the MiG-29 as a symbol of perseverance, valor, and sacrifice.

As the MiG-29's operational history draws to a close in many air forces, its legacy still endures. Russia has largely phased out the MiG-29 in favor of the Su-30 and Su-35, while Ukraine and the eastern NATO states have transitioned towards Western aircraft. Yet, the MiG-29 remains active in the air forces that still find value in its capabilities. It continues to serve in frontline roles, patrolling the skies and, in some cases, engaging in the last aerial battles of its long career.

The MiG-29's legacy is one of adaptation, survival, and dogged determination. The story of the MiG-29 is not merely about the aircraft itself, but the pilots who flew it; the nations that relied upon it; and the conflicts that have defined its place in history. As the last of these fighters gradually leave active duty, the Fulcrum's reputation will endure as one of the most iconic fighters of the Cold War era. Even as the world moves towards the next generation of tactical fighters, the MiG-29 remains a testament to the enduring spirit of combat aviation.

For every Fulcrum that once roared across the

skies; for every pilot who trusted his life to its capabilities, and for every conflict where it played a pivotal role, the MiG-29 will be remembered as a warplane that truly left its mark on history.

Select Bibliography

Badri-Maharaj, Sanjay. *Kargil 1999: South Asia's First Post-Nuclear Conflict*. Helion & Company, 2020.

Baker, David. *Mikoyan MiG-29 "Fulcrum" Manual: 1981 to present*. Haynes Publishing UK, 2017.

Brown, Craig. *Debrief: A Complete History of U.S. Aerial Engagements 1981 to the Present*. Schiffer Publishing, 2007.

Cooper, Tom. *Hot Skies Over Yemen: Aerial Warfare Over the Southern Arabian Peninsula: Volume 1; 1962-1994*. Helion & Company, 2017.

---. *Hot Skies Over Yemen: Aerial Warfare Over the Southern Arabian Peninsula: Volume 2; 1994-2017*. Helion & Company, 2018.

---. *War in Ukraine: Volume 6: The Air War February-March 2022*. Helion & Company, 2024.

---. *War in Ukraine Volume 7: Air War, January-December 2023*. Helion & Company, 2024.

Cooper, Tom, and Adrien Fontanellaz. *Ethiopian-Eritrean Wars: Volume 2 - Eritrean War of Independence, 1988-1991 and Badme War, 1998-2001*. Helion & Company, 2018.

Dimitrijevic, Bojan, and Jovica Draganić. *Operation Allied Force: Air War Over Serbia, 1999*. Helion & Company, 2021.

Gordon, Yefim, and Dmitriy Komissarov. *Mikoyan Mig-29 and Mig-35: Famous Russian Aircraft*. Crecy Publishing, 2019.

Groning, Andy. *The MiG-29: Russia's Legendary Air Superiority, and Multirole Fighter, 1977 to the Present.* Schiffer Publishing, 2018.

Guardia, Mike. *Skybreak: The 58th Fighter Squadron in Desert Storm.* Magnum Books, 2021.

Hallion, Richard. *Storm Over Iraq: Air Power and the Gulf War.* Smithsonian Books, 2015.

Henriksen, Dag, and Justin Bronk. *The Air War in Ukraine: The First Year of Conflict.* Taylor and Francis, 2024.

Lambeth, Benjamin S. *NATO's Air War for Kosovo: A Strategic and Operational Assessment.* Rand Corporation, 2001.

Wise, Alan R. *MiG-29 Flight Manual.* Schiffer Publishing, 2001.

Zuyev, Alexander, and Malcolm McConnell. *Fulcrum: A Top Gun Pilot's Escape from the Soviet Empire.* Grand Central Publishing (Hachette), 1993.

www.ingramcontent.com/pod-product-compliance
Lightning Source LLC
Chambersburg PA
CBHW041514090526
44751CB00010B/1113